REVISED

Nuffield Chemistry

TEACHERS' GUIDE I

ISBN 0 582 04633 5

General editor, Revised
Nuffield Chemistry:
Richard Ingle

Editor of this volume:
B. J. Stokes

Contributors:
E. H. Coulson
B. E. Dawson
A. M. Dempsey
H. F. Halliwell
Richard Ingle
M. J. W. Rogers
B. J. Stokes
G. Van Praagh
M. D. W. Vokins

REVISED NUFFIELD CHEMISTRY TEACHERS' GUIDE I

Published for the Nuffield Foundation
by Longman Group Limited

Longman Group Limited London
Associated companies, branches, and representatives throughout the world

First published 1966
Revised edition first published 1975

Design and art direction by Ivan and Robin Dodd

Filmset in Monophoto Times New Roman 327
by Keyspools Limited, Golborne, Lancashire
and made and printed in Great Britain
by Butler & Tanner Ltd, Frome and London

Contents

Foreword

It is now more than ten years since the Nuffield Foundation undertook to sponsor curriculum development in science. The subsequent projects can now be seen in retrospect as forerunners in a decade unparallelled for interest in teaching and learning not only in, but far beyond, the sciences. Their success is not to be measured simply by sales but by their undoubted influence and stimulus to discussion among teachers – both convinced and not-so-convinced. The examinations accompanying the schemes of study which have been developed with the ready cooperation of School Certificate Examination Boards have provoked change and have enabled teachers to realize more fully their objectives in the classroom and laboratory. But curriculum development must itself be continuously renewed if it is to encourage innovation and not be guilty of the very sins it sets out to avoid. The opportunities for local curriculum study have seldom been greater and the creation of Schools Council and Teachers' Centres have done much to contribute to discussion and participation of teachers in this work. It is these discussions which have enabled the Nuffield Foundation to take note of changing views, correct or change emphasis in the curriculum in science, and pay attention to current attitudes to school organization. As always, we have leaned on many, particularly those in the Association for Science Education who, through their writings, conversations, and contributions in other varied ways, have brought to our attention the needs of the practising teacher and the pupil in schools. This new edition of the Nuffield Chemistry materials draws heavily on the work of the editors and authors of the first edition to whom an immense debt is owed. The work leading to the first edition, published in 1966, was directed by Professor H. F. Halliwell, organizer of the Chemistry Project which carried out the trials in schools of the original draft materials. The editors of the first publication were:

H. F. Halliwell	*Introduction and Guide* *Book of Data*
M. J. W. Rogers	*The Sample Scheme Stages I and II: The Basic Course*
G. Van Praagh	*The Sample Scheme Stage III: A Course of Options* *Laboratory Investigations*
B. J. Stokes	*Collected Experiments*
E. H. Coulson	*Handbook for Teachers*
H. P. H. Oliver	*Background Books*

The new edition contains a preponderant part of these authors' material, either in its original form or in edited versions. They have all acted as consultants on the course the revision should take. They are credited among the authors of the new edition but their wider contribution in providing a firm basis for further developments must be gratefully acknowledged here.

I particularly wish to record our gratitude to Dr Richard Ingle, the General Editor of this new series. It has been his responsibility, together with Professor E. H. Coulson as consultant, to organize and coordinate this revision and it is largely through their efforts that we have been able to ensure the fullest cooperation between teachers and the authors. Thanks are due also to the Consultative Committee under the chairmanship, first, of the late Professor Sir Ronald Nyholm and, since his death, of Professor J. Millen, and especially to the group of practising teachers which met regularly to hear reports on the progress of the revision and to give their advice. This group consisted of:

H. S. Finlay
A. D. Gazard
J. A. Hunt
David J. Keeble
Michael Shayer
B. J. Stokes

As always I should like to acknowledge the work of William Anderson, our publications manager, his colleagues, and, of course, our publishers the Longman Group Ltd for their continued assistance in the publication of these books. I must also record our appreciation of those members of Penguin Education who worked on these books until a late stage in their preparation. The editorial and publishing contribution to the work of the projects is not only most valued but central to effective curriculum development.

K. W. Keohane
Coordinator of the Nuffield Foundation Science Teaching Project

Editor's Preface

The Nuffield Chemistry proposals described in this *Teachers' guide* were first published in 1966 as a contribution to the reappraisal of the place of science in education that was then taking place. The teaching schemes proposed were intended for children of average or above average ability in the 11–16 year age group who were expected to take an examination equivalent in standard to O-level, and were based on the Policy Statement of the Association for Science Education (1961) and the work of the Chemistry Panel of that Association. Since the proposals were first published they have been adopted by many teachers, and in this second edition a substantial revision has been undertaken in the light of the experience that has been gained. The principal points of revision are:
1. The range of publications for both teachers and pupils has been restructured; the new collection of books and other materials is described in Chapter 6 of this volume.
2. Although the main outlines of the teaching schemes have been retained, many amendments have been made to points of detail, and an alternative scheme is offered for Stage II.

This book is the first of three self-contained volumes of the new *Teachers' guide*. It is itself in three parts. In Part 1 the approach to chemistry teaching is discussed. This is followed by an account of the aims and objectives, and an outline of the content, of the teaching schemes offered for the first two years of the pupils' work in chemistry.

The second part of the book describes in detail two alternative schemes which may be followed during these first two years. They were first written as a result of the demands for more detailed information from those teachers who took part in the first trials of Nuffield Chemistry material. They found that so much in the approach was unfamiliar to them that they needed more help over such aspects as the best way to put across a particular point, the time that should be allowed for each Topic, the depth to which it should be pursued, and exact details of suggested experiments. The result is that these schemes probably contain more detail than many teachers require. In using this part of the book it is important to understand why it has been written in this form and the spirit in which it should be used.

Perhaps the most significant point is that the schemes are intended as samples only, written for the guidance of those who want them, but not intended to limit those who like to teach chemistry in this spirit while using their own schemes.

Three main groups of teachers have been in mind. Some will read Part 1 of this book and decide to produce their own scheme based on the principles outlined there. Others may produce their own scheme but using the main Topics of Part 2, perhaps in a different order, incorporating those sections of the materials that they find most useful. Yet others may prefer to accept one of the schemes as it stands, developing their own variations and amendments as they go along.

The alternative schemes offered are labelled A and B respectively. Readers will recognize that they differ from each other in flavour rather than in content.

Part 3 of this book consists of a series of appendices which discuss such general matters as laboratory organization, pupils' written work, examinations, and films.

The second volume of this *Teachers' guide* describes the teaching schemes proposed for Stage II and is also divided into three parts as described above. *Teachers' guide III* covers the options for Stage III. Each volume is complete in itself.

Part 1 Introduction

Chapter 1

A modern approach to chemistry teaching

The continuing need for curriculum development in the sciences

When the Nuffield chemistry proposals were first published, in 1966, they appeared at a time when there was much criticism in the teaching profession of both the methods and the aims of science teaching which were then generally current. Many teachers objected to the rigidity that came over their subjects when presented to pupils as a set of facts to memorize or a hard and fast system for discovering 'truths'. Many university teachers also brought the charge that their students came to them ignorant of modern scientific ideas and techniques, and with their critical and imaginative faculties undeveloped by their years at school. Furthermore, the need for more people to be aware of the social and economic aspects of science was becoming generally understood. It was widely felt that a training in science should lead to an understanding of what science is, and how it is practised, and that such training should form part of the education of all.

The response of chemistry teachers in the last few years, of which the Nuffield books were only a part, has taken much of the sting out of these criticisms. We cannot speak here of other reforms, but the very success of the Nuffield books and materials has demonstrated that their publication filled a deeply-felt need. It was always the hope of the original writers that their first books should be revised and brought into line with new ideas and needs as these arose. Fortunately the success of the first edition has enabled this revision to be put into effect. The new books incorporate a great deal of the first editions; but as a result of enquiry into the use and suitability of the materials in a large number of schools in which the proposals had been followed for the five years leading to O-level, it became clear that there were several areas where the understanding and response of the pupils could be improved. The first materials were, of course, based on a series of trials carried out in schools. We have been able to treat the first edition for our purposes as yet another trial, and our alterations have been made in the light of suggestions for improving the materials from teachers and pupils in schools.

The Nuffield proposals in outline

The main proposals on which the Nuffield chemistry schemes were based, and which remain unchanged in this revised edition, are four.
1. Pupils should gain an understanding that lasts throughout their lives of what it means to approach a problem scientifically. They should be taught to be aware of what scientists are doing and can do. This has little connection with the short-lived remembering of dictated information. Therefore science should be presented to pupils as a way in which *they* can conduct an enquiry into the nature of things as well as a body of information built up by the

enquiries of other people. Pupils must approach their studies through experiments designed to awaken the spirit of investigation. They must be given opportunities to observe and explore so that they develop disciplined imaginative thinking and are made fully conscious of the important part that science plays in modern life.
2. Which ideas are discussed, how they are presented, what materials and techniques are used to demonstrate them, depend on the level of development reached by the pupils and should be chosen accordingly. But what is taught must reflect up-to-date thought and technology.
3. Examination marks – whether in class tests, school, or public examinations – must be awarded chiefly for intellectual and manipulative skills, understanding, and a lively critical mind.
4. The teaching profession already possesses an imaginative outlook, wide experience, and a sympathetic understanding of pupils' needs and interests. If these essential qualities are to have their fullest effect, science teachers must not only be given freedom from restrictions that hinder these qualities, but also be encouraged to use and develop them. We especially hope that graduates thinking of becoming teachers will find, in the opportunities for individual and creative planning and in the research that is needed, a challenge that attracts them.

Where do these proposals lead?

If we accept these four proposals, where do they lead? They are based largely on what 'being scientific' means to a scientist – the application and the personal commitment involved; the importance of the disciplined guess or 'hunch' as well as logical argument; the feeling of exploration; and the readiness to make apparently unwarranted jumps while knowing how to check their validity. This activity of science is nothing new and, in fact, is embryonic in much of the activity of younger children. Many teachers have made full use of the original Sample Scheme to develop this approach and drawing on their experience we have tried to show in this new edition how the proposals can be put even more fully into practice. This *Teachers' guide* gives detailed suggestions for lines along which the spirit of enquiry can be developed. Teachers, pupils, and examiners are necessarily affected.

Teachers

The original proposals implied a far greater freedom to select contents and to plan our approach than was generally customary. Our job is still to see that the classroom offers pupils an opportunity for the kinds of experiment and observation we have in mind, and that, under our guidance, these are profitable and coordinated.

The 'new look' must go on provoking a 'new reaction'. We must continue to approach our job with fresh eyes. This is a formidable problem and its solution depends on that personal art which is the essence of teaching. Chapter 6 summarizes the Nuffield publications which have been designed to help. Here we will merely outline some of the principal problems.

1. Our chief concern will be to encourage pupils to be scientific about a problem. This means that they must have mental and manipulative skill in the exploration of a situation which, though familiar to us, is new to them. In all new situations one gropes and fumbles and is likely to make mistakes. This, however, is the exercise by which judgment develops. A pupil must have graded opportunities to be right or wrong, and he must be guided and encouraged to become better at finding out whether or not he is right. This is time-consuming at first but only time-wasting in the context of having to cover a traditional syllabus; properly organized, it brings considerable educational benefit later on. Our hardest task will be to extricate ourselves from 'the straitjacket of chronic success' and be willing to reconsider our methods. For example we have to learn to judge when to keep silent, leaving the pupil to puzzle the problem out by himself, and when to give encouragement and advice. That the pupil should see the point of experimental work is of the greatest importance. He must learn that chemicals, test-tubes, thermometers, and sources of heat and electrical energy are tools to which he should always turn to settle a point of curiosity or of speculation. He must learn that it is proper, not wrong, to have an opinion or an idea about something he observes, but he cannot begin to think he is scientific if he does not check whether his idea fits the observed experimental facts.

2. If we are to develop science teaching along the lines of these proposals, the appropriate organization of laboratory work for classes of thirty or more will play a very important part. Each experimental undertaking involves three *equally* important stages in which the pupil must take an active part:

a. Planning how to tackle the problem.
b. Carrying out the experimental work.
c. Discussing what deductions may, and may not, be made from the results.

Stages (*a*) and (*b*) need skilful guidance and encouragement by the teacher and must not be hurried. In order to have the time, practical techniques which are more streamlined than is customary will often need to be devised. Many teachers and classes have collaborated in the production of the Nuffield proposals, and the resulting suggestions for laboratory procedures are given in Part 2 of this book. One example will be given here.

An early experiment in Alternative B, given later in this book, requires the recrystallization of crude naphthalene from industrial methylated spirits. If the solid impurities are separated from the hot solution by filtration, under gravity, the experiment takes nearly the whole of a single period. If a centrifuge is used, however, beautiful clean, white, dry crystals can be obtained in less than fifteen minutes. There is then time available perhaps to follow up suggestions and to improve the process and certainly to discuss and talk around the experiment.

3. The introduction of theory – of molecules and ions, and of the writing of equations – raises a major problem in planning a chemistry course. This is discussed in the *Handbook for teachers*. Here we will just say that speculation about material changes in terms of molecules, ions, and atoms is the very essence of chemistry and we introduce it at an early point in our proposals. From the point of view of its effect on the pupils, the topic is full of the danger of unscientific dogmatic assertions. The ideas associated with any model which is introduced should be strictly limited to those necessary to deal with the questions arising from the pupils' experiments. In whatever way this problem is eventually solved, we want pupils to learn to distinguish between observed phenomena and explanations put forward by the creative thinking of the human mind. Pupils must learn to see the interplay between observed fact and explanation, as summarized in the following diagram, and to appreciate how science develops through this interplay.

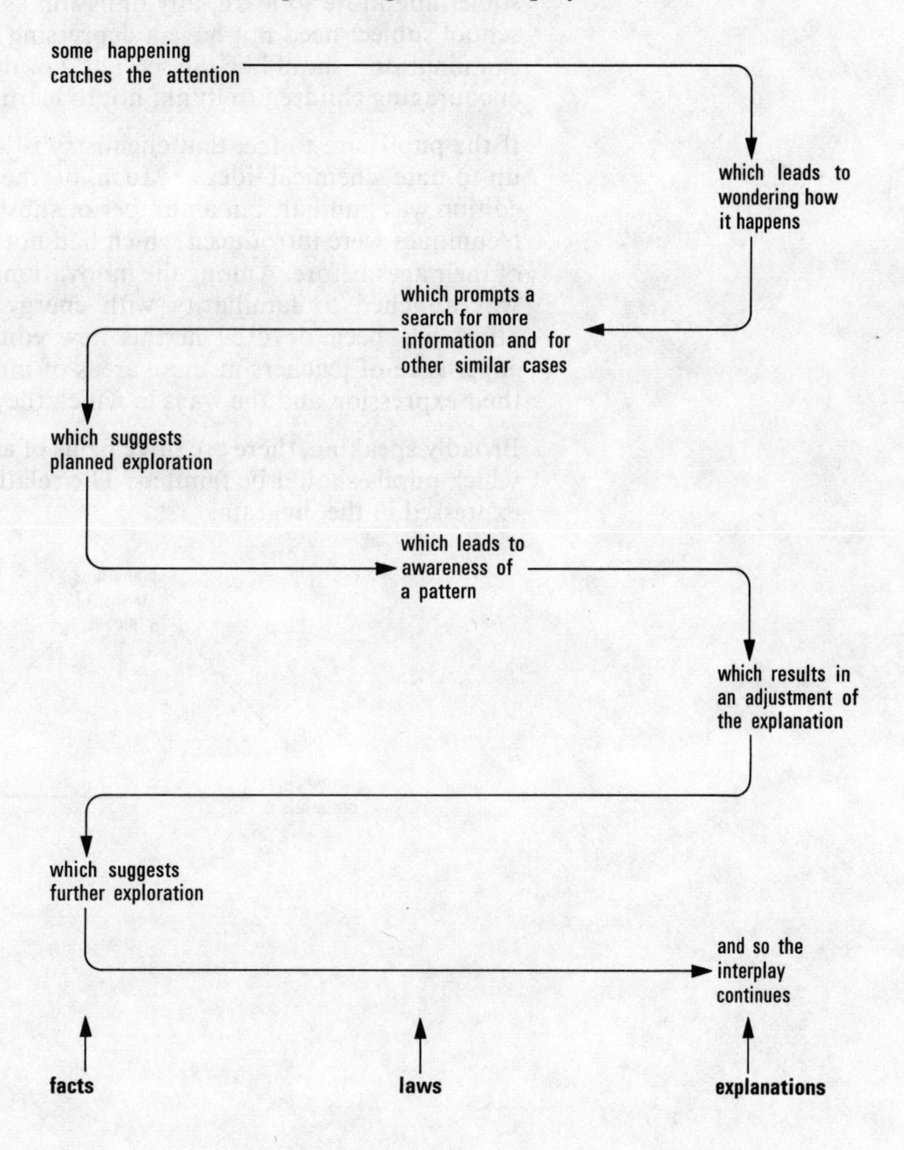

In almost every lesson pupils must be given opportunities to take an active part in this kind of interplay. From the beginning they must be encouraged to think imaginatively about the problems confronting them, and to suggest further experiments and explanations. When they are young, many children will do this without prompting and we want this tendency to grow in precision and not to be stifled because children are afraid of being 'wrong'. *We* can distinguish between the flippant and the serious and, if we want pupils to develop disciplined creative thinking, we must develop the necessary classroom techniques.

4. What should be done about the content of a course? Many present courses suffer from excessive loading and this has become serious. We, however, see this problem as arising from an attitude that regards a school subject as a body of information which pupils should be able to repeat. As a result, any addition is regarded as something more to learn, any omission as a loss of knowledge. A school subject need not have a depressing effect, and content, like examinations, should be the servant, not the master. We should be encouraging children to think, not to learn jargon.

If the pupils are to feel that chemistry is 'alive', they should meet up-to-date chemical ideas. Much of the chemistry in the first edition was familiar, but a number of substances and experimental techniques were introduced which had not been taught to children of their ages before. Among the innovations, particular importance was attached to familiarity with energy changes. Considerable effort has been devoted in this new edition to drawing on the experience of teachers in these areas of innovation and in refining their expression and the ways in which they can be taught.

Broadly speaking, there are three fields of activity in chemistry with which pupils should be familiar. The relationship between these is expressed in the diagram.

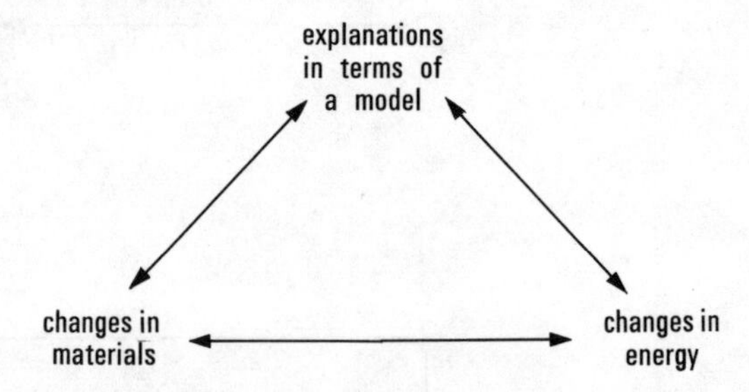

But if the Nuffield proposals make demands on teachers, they also make demands on pupils and examiners.

Pupils

Over the five years between the ages of 11 and 16 the pupils mature rapidly. In the first year or two they get the feel of what being a scientist means, meet new and strange materials, learn how to do a job, enjoy the thrill of discovery, and begin to speak the language. It is not a time for being forced to learn by heart, but a time to collect experiences, to collect ideas, to collect information, to explore new territories. Pupils develop a skill in doing experiments and asking questions. As they get older, this exploration becomes more deliberate. Attempts at explanations in terms of atoms, ions, and molecules become second nature, but pupils must realize that their explanations are tentative and must be tested and, if need be, adjusted, or even rejected.

During the fifth and final year of the course, their interests and abilities have become clear enough to show that they need different areas of exploration. The extent to which each pupil carries his questions will also vary.

Examiners

Any educational scheme will be made or broken according to whether or not, at times of assessment, in public examinations or at school, the demands made by the types of questions set encourage the intentions behind the scheme. This fundamental problem is discussed in more detail in Appendix 5, towards the end of this book.

Chapter 2

The form and content of Nuffield Chemistry

What is expected of the pupil?

Many chemistry syllabuses, whatever may have been the original intention of their compilers, are predominantly lists of specific materials and concepts. The Nuffield Chemistry proposals have led to the formulation of a syllabus emphasizing manipulative and intellectual skills, and its requirements are stated in terms of abilities rather than of items of information that must be remembered. Nevertheless, the actual chemical content through which pupils attain this manipulative and intellectual nimbleness and understanding, and through which they acquire the habit of demanding and assessing evidence for statements, has often been the familiar material used in more traditional syllabuses.

The first enduring lesson to be learned is that chemistry is a way of conducting an enquiry. As a result of investigations and problems undertaken during the course, pupils should have acquired skill in, and understanding of, the following:

1. *Getting new materials from those available*
Bringing about changes: use of thermal and electrical energy and energy from electromagnetic radiation; adjusting the conditions of the experiment (such as temperature and concentration); the effects of adjustments on how fast the reaction takes place and how far it goes.

Knowing whether or not something new has been obtained – using separation techniques such as fractional distillation, crystallization from water and other solvents, centrifuging, and chromatography.

Using a melting point or a boiling point to test the purity of the product and using simple chemical means of finding out in what way the new products are different.

2. *Looking for a pattern in the behaviour of substances*
Representing and interpreting qualitative and quantitative results, and assessing the evidence for the interpretations. Familiarity with major patterns of behaviour – examples of this are acidity and alkalinity in aqueous solutions; metallic and non-metallic nature; electrical differentiation into electrolytes and non-electrolytes; colloidal and non-colloidal systems; the periodicity of properties of elements and the conservation of mass and energy. Realizing that the importance attached to a pattern changes in the course of time – for example, Lavoisier's, Davy's, Arrhenius's, and Brønsted's ways of thinking about acids. The discussion should recognize that a continuous gradation of properties exists in all classifications.

3. *Using explanatory concepts and knowing how to check theory by observation*
Using the ideas of atoms, ions, molecules, giant structures, and types of bonding to explain experimental observations; planning experiments to test theoretical explanations; using the concept of the mole as an aid in this testing; the beginnings of deliberate molecular engineering in terms of chains, rings, sheets, and three-dimensional structures.

4. *Associating energy changes with material changes*
Measuring the 'heat' of a reaction in simple ways; using tables of standard heats of formation; using energy changes in a simple way to check theory against fact by associating them with changes in energy due to bond making and bond breaking and with energies of vibration, rotation, and translation of ions and molecules.

Simple investigations of material changes brought about by absorption or emission of ultra-violet, visible, and infra-red radiation, for example photosynthesis, photography, and flame tests.

Getting an overall picture that chemical changes in raw materials can be used to supply energy, for example for heating; and that energy can be used to obtain new products from raw materials.

5. *Chemistry as an enquiry*
But chemistry is also the result of enquiry. Pupils should be aware of the work of other people. This does not mean a course of potted biography, but the opportunity to meet a few chosen examples of exciting minds and the flash of genius. Chemistry should be recognized as a product of people's activity, and the subject seen in a human perspective. For example, pupils should have some awareness of the way that the concept of an element, the concept of atoms, and the concept of energy have developed.

6. *The social and economic aspects of chemistry*
An appreciation, through a few specific examples, of the way the knowledge acquired by chemists is used for meeting and fostering some of the material requirements of a community.

Some awareness of the need to conserve natural resources and to avoid polluting the environment by the careless disposal of waste materials.

Lastly, at the end of the course, exploring the behaviour of substances should be linked with other scientific studies so that science is seen in relation to the life of the community as a whole – not only in the sense of direct application but also as an addition to a body of experiences which greatly affects our lives and our sense of values.

Within this framework, considerable choice is left to the teacher to adjust the course to the interests and abilities of the pupils.

The content of the five-year course

The detailed schemes which are given in the volumes of this *Teachers' guide* can be put together to give a Topic-by-Topic course lasting for five years. They are based on the approach to chemistry discussed earlier in this book, and have been used in many schools, in some cases for a number of years. The results have shown that the Nuffield Chemistry proposals can be translated into classroom practice to give a course that is both stimulating and worth while, but it is emphasized that the schemes are offered only as examples. Many teachers, especially those new to the Nuffield approach, will want to begin by using the schemes as they stand, but it is hoped that many of them will take advantage of the flexibility that this type of course offers, and devise schemes of their own to suit their own requirements. The five-year course is divided into three stages, which will now be discussed separately.

Stage I *Exploration of materials*

During this stage the child will learn about a wide range of materials, how to separate them, and about some of their patterns of behaviour. Thus he is led up to realize the central part played by the elements in chemistry. Two years have been allowed for this but some teachers may wish to move on to Stage II a term earlier. However, the child's progress during this stage must not be too hurried. It is a time for asking questions and for getting help in finding out answers. Speculation at this point is about the behaviour of materials and the ways of succeeding in one's experimental aims. In this respect Stage I has a different purpose from that of Stage II where this type of speculation is broadened to consider the particles which are imagined to be responsible for the behaviour of the material on the bench. Nevertheless there should be no sharp division in the pupil's mind between Stage I and Stage II. The time when pupils need to pass on to Stage II is when, because of what they have learned in Stage I, they begin to be increasingly aware of the complicated structure of materials and the orderliness of their behaviour.

Stage II *Using ideas about atoms and particles*

The word here on which most emphasis is placed is 'using'. The ideas are regarded as tools to be used, perhaps clumsily at first, but with greater precision and discernment as time goes on. The essential focus of chemistry must be on materials and the way they behave. 'Models' of atoms, molecules, and ions, whether in expanded plastic or in the mind's eye or in the form of chemical equations, are used to aid understanding and in turn to focus attention even more sharply on the test-tube. Just like tools, these ideas need continual overhaul and sharpening (and sometimes replacing by something more effective).

Molecules, giant structures, ions, and three-dimensional structures have therefore been introduced as soon as the pupils have enough experience to consider them reasonable, and as soon as they are useful. Therefore these concepts are not kept back until their acceptance can be argued more rigorously, but they are introduced in order that the use to which they are put will lead to critical understanding. In doing this we hope we are acting in the spirit of the advice of a great nineteenth-century teacher:

> 'After many trials in the course of my teaching, I have come to the conclusion that, not only is it impossible to eliminate the atomic theory altogether, but, moreover, in order to arrive at this theory, it is not desirable to follow the long and fatiguing road of induction. On the contrary, it is better to get to it as quickly as possible, by one of those short cuts which the human mind often takes in order to raise itself quickly to a height from which the relations between phenomena can be discerned at a glance.'
> *S. Cannizzaro (1861).*

At the same time we also align ourselves with Cannizzaro when he said, three years earlier, that he was not one of the teachers who:

> 'allowing themselves to be carried away by the desire of being brief and concise, and of rendering the approaches of our science accessible even to the least prepared intellects, explain dogmatically all the laws and theories in a synthetic order, concealing their origin and development. . . .'

Stage II occupies a little over two years. It includes Topics that bring the pupils' attention to the way that chemistry can be deliberately directed to meeting some of man's social and industrial needs, providing there is added the experience and insight of the engineer and technologist. Their attention has in fact been directed to these needs on many occasions in earlier parts of the course, particularly through the use of the specially designed film loops, and this is part of the wider intentions of the chemistry proposals. At this stage it is done more explicitly but in a way that allows flexibility of treatment.

Stage III *A course of options*

Stage III can be seen as an elaboration of the type of work that has been done by many chemistry teachers once O-level examinations are over and the restrictive requirements of public examinations are lifted for a brief period. The aim of the chemistry proposals is to develop manipulative and intellectual skills within the framework of chemistry: the last stage is therefore used to give pupils opportunity to round off these skills and bring themselves to a reasonable standard of lively competence. By this time in a school course, however, there will be a greater differentiation of ability and interest among pupils. Not all will benefit by exercising themselves in the same prescribed area: a variety of topics has therefore been suggested and it is left to the teachers to judge which of these will best suit the needs of their pupils.

Time allocation In planning the amount of work to be done, the allocation of teaching time adopted has been that recommended in the Policy Statement of the Association for Science Education, published in 1961:

Years 1 and 2 (Stage I)	1 double period in a laboratory and 1 homework period per week.
Years 3, 4, and 5 (Stages II and III)	1 double and 1 single period in a laboratory and 1 homework period per week.

A school teaching year has been taken to consist of 30 weeks.

Chapter 3

The aims and objectives of Stage I

General aims

The Nuffield proposals described in Chapter 1 form a set of general aims to be achieved in the teaching of chemistry. These aims are not all equally appropriate for all Stages of the course. In Stage I, most pupils will need plenty of direct experience of phenomena if their learning is to be meaningful, and their ability to understand abstract ideas will only just be emerging. As was said earlier, at this Stage pupils should be meeting new and strange materials, and beginning to speak the language of chemistry. It is a time for new experiences and ideas, and not simply a time for learning by heart. The necessity for keeping records should not be overemphasized. Young pupils want to be active most of the time, and should not be hampered by too much written work, which they often find difficult, and at which most of them are very slow. They should be encouraged to regard their notebooks as diaries in which they record the more important things they do, for the building up of a notebook is in most cases enjoyable and stimulating if it is personal. With this appreciation of the needs of young pupils a set of aims for Stage I can be produced. The pupils should:

1. Acquire basic knowledge about the behaviour of substances.

2. Seek patterns in the behaviour of substances.

3. Develop concepts related to the classification of substances and their observable behaviour.

4. Develop manipulative skill in laboratory procedures with common apparatus.

5. Gain confidence in learning by discovery.

6. Record their work in a personal style and an accurate manner.

7. Find out about the sources and uses of the substances they meet, and any hazards connected with them.

Objectives and assessment

The general aims stated in the previous section are broad statements of intent. We may distinguish such aims from *objectives*, which give a clearer picture of what a pupil should have understood, or be able to do, after pursuing a piece of work. In this book we shall distinguish between *general objectives*, which are discussed in this section, and the *objectives for pupils* which are given at the start of each section of Part 2 of this book. An example may make the distinction clear. *General aim* 1, given above, is to 'Acquire basic knowledge about the behaviour of substances'. The related *general objective* for the course as a whole might therefore be 'Facility in recalling information and experience', and one of the *objectives for pupils* for a particular section of Part 2 might be 'Knowledge of the industrial process of crude oil fractionation'. The pupil may therefore *aim* at acquiring basic knowledge; if he has achieved this aim he should be able to remember the knowledge (a *general objective*); and in one particular section the information to be able to remember is the industrial fractionation of crude oil (an *objective*).

The general objectives of the Nuffield proposals now follow. Of these, the first five are the most important at Stage I level. The remaining three are of lesser importance at this level, but they should not be overlooked. They will have some part to play in Stage I, but will be of greater importance in Stage II.

1. Facility in recalling information and experience.

2. Skill in handling materials, manipulating apparatus, carrying out instructions for experiments, and making accurate observations.

3. Skill in devising an appropriate scheme and apparatus for solving a practical problem.

4. Skill in handling and classifying given information (including, in Stage II, graphical representations and quantitative results).

5. Ability to interpret information with evidence of judgment and assessment.

6. Ability to apply previous understanding to new situations and to show creative thought.

7. Competence in reporting on, commenting on, and discussing matters of simple chemical interest.

8. Awareness of the place of chemistry among other school subjects and in the world at large.

It will be seen from Appendix 5, on examinations, that it is these general objectives that must be borne in mind when putting together examination papers to assess the progress of the pupils.

The more detailed *objectives for pupils*, given at the beginning of each section of Part 2, are offered in the hope that they will help to focus attention on what is required of the pupil in that particular section. Like other aspects of the Nuffield schemes they must not be regarded as invariable; lessons may often go in ways other than those described in this book and achieve other equally worthwhile objectives. Indeed, these objectives must be varied by teachers who wish to teach according to the Nuffield proposals but develop their own schemes to achieve the same general aims.

Chapter 4

The content of Stage I

Two alternative schemes are offered for Stage I, labelled as A and B respectively. Each scheme is divided into ten Topics, the Topics in turn being divided into a number of sections. An outline of the contents of each Topic now follows. At the end of the chapter the alternatives are compared.

An outline of Alternative A

Topic A1 *Getting pure substances from the world around us*

A1.1 How can we get pure salt from rock salt?
A1.2 Is ink a single substance?
A1.3 What coloured substances can we get from grass?
A1.4 Using coloured substances from plants to provide an introduction to acidity–alkalinity
A1.5 What can we get from crude oil?
A1.6 How can metals be obtained from rocks?

The idea of the individuality of substances is something that can only grow with experience. At first the words 'pure substance' will have little meaning, but the pupils will have met such phrases as 'pure table salt' and 'guaranteed pure'. This Topic, which may last nearly a term, has three purposes:

1. To give the pupils a feeling for the individuality of different materials, so that they feel the need for the terms 'pure substance' and 'mixture'. Any precise definition cannot be made until much later, and should not be attempted.

2. To enable the pupils to use some of the tools and techniques of chemistry: how to dissolve, what solvents to use, how to filter, evaporate, and distil (using the simplest apparatus), how to separate by paper chromatography, and how to use electricity to get new materials.

3. To show the pupils that what they are doing is similar to what is done in industry, but there it is done on a large scale. Here the film loops are an integral part of the course. They are not intended to be 'learned', but to act as brief intervals of interest and stimulation which take the lesson into the world of industry and engineering.

After this Topic the pupils should feel more and more manipulatively competent in the business of getting to know about the materials of the world, and should have some idea of their remarkable diversity.

Topic A2 *The effects of heating substances*

A2.1 What happens when substances are heated?
A2.2 A further look at the effect of heating copper sulphate crystals
A2.3 To find out if there is any change in weight when substances are heated
A2.4 Tracking down the matter lost when potassium permanganate is heated
A2.5 Tracking down the matter gained when copper is heated

By the end of this Topic a pupil should have met quite a variety of materials and become aware of the fact that they can be changed into new materials. The basis of reference for the next Topic is laid and the balance begins to be used as a tool for checking whether everything has been accounted for. It is a period when the keeping of a diary of investigations needs time for guidance. Brighter pupils will work more quickly but may well find more problems on the way.

Topic A3 *Finding out more about the air*

A3.1 Is air used up when copper is heated in it?
A3.2 The history of the discovery of oxygen

This Topic takes about three weeks. It follows on from what was met in Topic 2, and it enlarges the pupil's experience by making him deal with gases. The content of this Topic is traditional but the treatment is not, and there is opportunity in A3.1 for making suggestions and trying them out. By the end of the work the pupils should have had some experience in handling gases and they should know that air is a mixture, chiefly of nitrogen and oxygen, and should also be able to talk fairly competently about the evidence for this.

Topic A4 *The problem of burning*

A4.1 What are the properties of the compounds formed in burning?
A4.2 Do other substances that burn use oxygen?
A4.3 Investigating changes in weight when a candle burns

After two terms or more of work, pupils should have a fairly sharp awareness that the stuffs of this world are made up of individual materials. These can be separated, put into bottles, given a name, and recognized when met again by the way they behave. Pupils, therefore, should be well on the way to understanding the use of the words 'pure substance'.

Topic A4, and possibly A5, round off the first year's work with investigations and experiences which develop a new realization – that pure substances are made up from, and must somehow contain, other substances. This, of course, is not an idea with which the pupils begin the Topic – they are looking further into this business of burning – but they should have arrived at it by the end of the time suggested for Topic A4. There is a use for a different label, and the word 'compound' as distinct from 'mixture' summarizes both our awareness and our ignorance. Nevertheless the words can be used, and pupils can learn how to judge whether they are using them justifiably; but, again, the sharpness of this judgment will have to be improved by use, before a definition is worth while or possible. The second step, which leads into Topic A5, is then not difficult. Some pure substances seem 'made up' only of themselves. These are the building bricks of the world and they are obviously sufficiently important to be worthy of a special label – elements.

Topic A5 *The elements*

A5.1 What is an element?

A5.2 Into what groups can we sort elements?

Although only about three weeks are suggested for this Topic, it ushers in a period of considerable hunting and collecting by the pupil. The exploration of the world of elements involves the library even more than the laboratory. Anything pupils feel moved to find out about *any* element is of value because of the activity that brought it to light. They should be encouraged to compile their own booklets about the elements. This occasion is full of opportunities for simple diagrammatic representation of information.

Topic A6 *Competition among the elements*

A6.1 Putting elements in order of their reactivity with oxygen

A6.2 Where does carbon come in the series?

This Topic, for which a maximum of four weeks are suggested, follows up a combination of earlier ideas (A1.6 and A5). By the end of it, some feeling for 'order of reactivity' should be established.

Topic A7 *Water as a product of burning*

A7.1 What is the liquid condensed from a Bunsen flame?

A7.2 Is water an oxide?

A7.3 What other metals can we use to obtain the inflammable gas from water?

A7.4 A more convenient way of preparing hydrogen in the laboratory

A7.5 Can hydrogen be used to reduce metal oxides?

This is the first place in the course where a particular material has been chosen for extended investigation. At least half a term is suggested for this, but the work breaks up into fairly self-contained sections. It is also a Topic that should be more widely treated than is usually done in a chemistry course. There should be a link with geography (rivers, the rain cycle, erosion, and the sea – the latter being checked back with A1.1 and forward with A10), with biology, and with physics.

By the end of this Topic a pupil should know that water is a compound of hydrogen and oxygen and be able to talk clearly and simply of the experimental evidence for this. He should also have become fairly familiar with the new element, hydrogen, seeing it in the light of his previous experience of competition among elements. He should have met the fact that hydrogen is, volume for volume, the lightest known substance; later it will be reasonable to take hydrogen as 'element number 1'.

Topic A8 *The effects of electricity on substances*

A8.1 Which substances conduct electricity?
A8.2 Do solutions conduct electricity?

The fact that electricity is useful for getting new materials has already been met in A1.6, and will be met again in A10. In this Topic, which should last about five weeks, pupils will realize that the Bunsen burner and the battery are the two major tools for getting new materials. It is a time when a new technical area has to be explored and there must be a link with the work that the pupils do in physics. By the end of the Topic pupils should know how to set up an appropriate circuit and should realize that metals can be expected at the electrode attached to the negative terminal of the accumulator or battery, and non-metals at that attached to the positive terminal. They should be able to *use* the words anode, cathode, and electrolyte with confidence and should have the beginnings of a feeling for using either a Bunsen burner or a battery in a new situation.

Topic A9 *Chemicals from the rocks*

A9.1 How can iron be extracted from iron ore?
A9.2 Investigation of a mineral: malachite
A9.3 Investigation of limestone

This Topic and the next, which take about five or six weeks each, round off Stage I, Alternative A. The somewhat haphazard exploration of the first year is becoming more and more a deliberate investigation. The pupils have quite a varied experience to call upon, as they become familiar with the ideas of elements and of degrees of chemical vigour, and have a range of experimental skills. Film loops will take the work pupils are doing in the school laboratory out into the world of industry and the engineer.

Topic A10 *Chemicals from the sea*

A10.1 What chemicals can be obtained from the sea?
A10.2 What is the effect of electrolysing sea water?
A10.3 Getting iodine from seaweed
A10.4 A family of elements – the halogens
A10.5 The industrial production of the halogens

This Topic continues the type of work involved in A9 but it involves the use of electricity and it offers further experience of a family of useful elements – the halogens. A range of film loops is available.

An outline of Alternative B

Topic B1 *Separating pure substances from common materials*

B1.1 How can crude alum be purified?
B1.2 How can crude naphthalene be purified?
B1.3 What is the difference between tap water and distilled water?
B1.4 How can water from the sea be purified?
B1.5 What coloured substances can we get from grass?

The idea of the individuality of substances is something that can only grow with experience. At first the words 'pure substance' will have little meaning, but the pupils will have met such phrases as 'pure table salt' and 'guaranteed pure'. This Topic, which may last nearly a term, has three purposes:

1. To give the pupils a feeling for the individuality of different materials, so that they feel the need for the terms 'pure substance' and 'mixture'. Any precise definition cannot be made until much later, and should not be attempted.
2. To enable the pupils to use some of the tools and techniques of chemistry: how to dissolve, what solvents to use, how to filter, evaporate, and distil (using the simplest apparatus), how to separate by paper chromatography, and how to use electricity to get new materials.
3. To show the pupils that what they are doing is similar to what is done in industry, but there it is done on a large scale. Here the film loops are an integral part of the course. They are not intended to be 'learned', but to act as brief intervals of interest and stimulation which take the lesson into the world of industry and engineering.

After this Topic the pupils should feel more and more manipulatively competent in the business of getting to know about the materials of the world, and should have some idea of their remarkable diversity.

Topic B2 *Acidity and alkalinity*

B2.1 Ways of detecting acidity
B2.2 How can acidity be cured?

This Topic takes at most four weeks. How acid or how alkaline solutions are (or whether they are neutral) is found by matching colours and reporting by a scale of numbers – pH scale. There is no discussion of categories of materials called acids or alkalis. This work obviously links with some aspects of rural science and biology, but it is developed by reference to common, domestic, and everyday substances.

Topic B3 *Fractional distillation as a way of separating mixtures*

B3.1 How can a mixture of two liquids be separated?
B3.2 Fractional distillation of crude oil
B3.3 Fractional distillation of air

As a result of the three or four weeks spent on this Topic pupils should have learned how to carry out a distillation in a simple way, and have again seen that a boiling point is a useful distinctive property. Again the film loops form an integral part of the lessons bringing out the similarity between what is done in class and what is done in industry. The Topic ends by reversing this use of the film loop, in that the fractional distillation of air provides evidence for the approximate composition of air.

Topic B4 *The major gases of the air*

B4.1 What are the properties of oxygen and nitrogen?

This is the first experience of exploration of new materials during Alternative B and, because gases are involved, it brings out the necessity of new techniques in transference and handling. About three weeks are suggested for this Topic, but the way the time is spent will depend a great deal on the pupils' previous knowledge. By the end of the time pupils should know:

1. Something about how to handle gases.
2. That oxygen is the more reactive constituent of the air as far as burning is concerned.
3. That when materials burn in the air new substances are formed and heat is given out. Whether we are interested in the former or the latter or in both depends on what we want to do.
4. That these new substances have their own peculiarities as is seen in their appearance and in the pH of their solutions in water.

This Topic is only a first attack on the problem. It is followed up again in more detail two terms later (see B9).

Topic B5 *Finding out more about substances by heating them*

B5.1 What happens when substances are heated?

B5.2 A further look at the effect of heating copper sulphate crystals

Over a period of four weeks we look further into what happens to materials on heating them. The question whether the air always plays a part leads to the introduction of weighing as a means of checking. The Topic involves not only experience with a variety of materials, but also a more careful and detailed study of one example – heating blue copper sulphate crystals. By the end of this Topic pupils should have gained:

1. A fairly clear idea of the differences between heating and burning.
2. Their first taste of a planned investigation which involves argument and discussion, and the use of many of the techniques already met.
3. The realization that water is a chemical, and how to recognize it.

Topic B6 *Using electricity to decompose substances*

B6.1 Investigation into substances which conduct electricity

B6.2 Further investigation into substances which conduct electricity

B6.3 What happens when solutions conduct electricity?

B6.4 Using electricity for plating

B6.5 Electricity from chemical reactions

Here is another way to get new materials that lends itself to a great deal of profitable fun. By the end of the Topic pupils should know how to set up an appropriate circuit. They should also realize that metals can be expected at the electrode attached to the negative terminal of the accumulator, and that other materials can be expected at the electrode attached to the positive terminal. There should be liaison with the physics teaching here; this does not necessarily mean that there should be concurrency of topics, but there should be close understanding of what is being done, and when.

Topic B7 *The elements*

B7.1 What is an element?

This is a brief, summarizing Topic which pulls together certain ideas that have developed up to now and forms the basis for the rest of Stage I.

Topic B8 *Further reactions between elements*

B8.1 Studying the difference in properties between elements and their compounds: hydrogen, oxygen, and water

B8.2 Studying the difference in properties between elements and their compounds: some sulphides, oxides, and chlorides

About six weeks are recommended for this opportunity to become familiar with the elements as such.

The Topic begins a period of considerable hunting and collecting by the pupil. The exploration of the world of elements involves the library even more than the laboratory. Anything they feel moved to find out about *any* element is of value because of the activity that brought it to light. They should be encouraged to compile their own booklets about the elements and the occasion is full of opportunity for simple diagrammatic representation of information.

From here on stress should be laid on the great differences to be seen between elements on the one hand and the compounds they make on the other hand.

Topic B9 *Investigation of some common processes involving the air*

B9.1 What chemical changes take place in burning?

B9.2 What chemical changes take place in breathing?

B9.3 What chemical changes take place in rusting?

B9.4 What is the chemical composition of rust?

Now that a more precise understanding of the use of the word element is established it is possible to investigate again, and more deliberately, the problems of burning and of what seem at first sight to be similar processes, namely breathing and rusting. About six weeks are suggested for this Topic.

Topic B10 *Competition among the elements*

B10.1 How can we get metals from their oxides?

B10.2 How metals are obtained from their ores

B10.3 Can metals be displaced from solutions of their salts?

B10.4 Can non-metals be displaced from solutions?

B10.5 The family of halogens and their industrial production

In this Topic pupils spend some eight or ten weeks in investigations from which they can arrive at two further generalizations:
1. Some elements show family resemblances and lists can be made, for example, of alkali metals, coinage metals, and halogens.
2. There is some kind of order of reactivity.

Again, the film loops will take the work that the pupils are doing in the school laboratory out into the world of industry.

A comparison of Alternatives A and B

The reader will already have seen that the two alternative schemes differ more in flavour than in content. The alternatives may be taught as they stand, or one or more Topics of one alternative may be used in place of the corresponding Topics of the other. The following comparison may help when deciding which scheme to use.

The aim of Topic A1 is the same as that of Topic B1. The time suggested is the same, about ten weeks, but the materials used are different. These Topics are therefore interchangeable. Acidity–alkalinity is developed more fully in Alternative B, where it is the subject of a complete Topic (B2, four weeks); the subject is introduced in Alternative A in section A1.4.

The two approaches to the composition of the air are both investigatory but quite different. In Alternative A oxygen and nitrogen are found to be components of the air after an experimental study of the effect of heat on substances, whereas in Alternative B the fractional distillation of liquid air (introduced by a film loop on the subject) leads to the two principal gases, whose properties are then investigated. The effect of heat on substances is in consequence dealt with less fully in Alternative B (Topic B5).

The idea of elements is met with in Alternative A at Topic A5, somewhat earlier than in Alternative B. Topics B7 and B8, both concerned with elements, may be seen as an alternative to Topic A5, but the examples chosen are different.

The use of electricity in getting new materials is met rather earlier in Alternative B (Topic B6) than in Alternative A (Topic A8). In the Alternative A approach the idea of elements comes before electrolysis, and is used to help the understanding of the latter, whereas the reverse is true of Alternative B; here electrolysis precedes the idea of elements and is used to help to define them. Topic B6 also includes an elementary consideration of simple cells, not mentioned in Alternative A.

Burning is dealt with towards the end of Alternative B, where it is considered in relation to breathing and rusting. In Alternative A it is investigated at a much earlier point (Topic A4).

Both Alternatives end with Topics in which pupils are led to investigations from which they can arrive at the ideas:
1. Some elements show family resemblances – the halogens are used to illustrate this point.
2. There is some kind of 'order of reactivity'.

Chapter 5

Background to the work of revision

The main curriculum materials which were the result of the development activities of numerous teachers throughout the country were first published in 1966. By 1970 about two hundred schools were entering some 10 000 candidates for the Nuffield O-level examination. By that time many teachers in these schools had, therefore, experience of teaching the whole Nuffield scheme and were in a position to give advice on how the publications should be revised. In 1970, the Nuffield Foundation decided that the time had come for an enquiry into the need for a revision, and into what form any such revision should take. During the academic year 1970/71 Dr Richard Ingle undertook this enquiry with the assistance of six teachers who formed a small consultative committee. Discussions first took place in many parts of the country with chemistry teachers and with other individuals who were involved in chemistry education. Chemistry teachers in the two hundred schools who had entered candidates for the 1970 examination were then invited to comment in detail by means of a questionnaire, on the need for revision. Finally, as many of these schools as possible were visited. The degree of interest in the work of revision both at home and overseas, combined with the increasing number of schools making use of the Nuffield Chemistry materials, indicated that the time was indeed ripe for a revision. This was put in hand in September 1971, under the full-time coordination of Dr Ingle at the Centre for Science Education, Chelsea College, London.

The main findings relevant to the revision are given below.

Suitability of the Sample Scheme for pupils

The whole question of the appropriateness of the subject matter of the Sample Scheme has been looked at very carefully from a number of points of view. Criticisms of the course have described it, for example, as 'too difficult except for very able pupils'. On the other hand it has been widely adopted in a great variety of schools and it was interesting to find out why it has had this particular success.

Stage I of the course has been very well received by teachers throughout the country and, even before the publication of *Nuffield Combined Science*, many teachers were successfully adopting it for use with a much wider range of ability of pupils than that for which Stage I was originally designed. Not only has the content been welcomed, but the suggested teaching method (which makes extensive use of laboratory-based investigations by pupils) has also been generally welcomed at this level. There are now two published versions – Stage IA and Stage IB – or three if we include Nuffield Combined Science. Since one or other of these three schemes appears to be satisfying the needs of many schools, it has not been thought necessary to carry out many major changes in the content of Stage I. There are, however, many changes to points of detail which have been made in the light of comments from schools.

As far as Stage II is concerned, the point of overriding importance which emerged from the enquiry was that, with a number of exceptions noted below, the course was thought by most teachers to be of about the right length and difficulty for pupils likely to take an examination at 16+ in chemistry. Furthermore, the inclusion of material on subjects such as structure, energy, rates, equilibrium, and radioactivity, which were considered revolutionary in O-level courses in the mid 1960s, is now generally accepted as desirable. Indeed it has been also largely accepted by those teachers who do *not* believe, at this level, in using so much laboratory-based investigation by pupils as is proposed in the Sample Scheme.

The main ideas which many teachers have reported their pupils finding difficult are the gram-atom (Topic 11) and ΔG (Topic 23). The work of Topic 11 has often been a serious stumbling block and teachers have pointed out how sharply the level of difficulty of the course rises at the beginning of Stage II. This is one reason why in *Teachers' guide II* a second sample scheme is provided (Stage IIB) in which the more difficult ideas are introduced more gradually, particularly in the first year of Stage II.

The existence of a number of suggested alternative teaching approaches in *Teachers' guide II* may suggest that the amount of work included in the basic course has been increased in the revision. Close inspection will show that this is not the case. In particular, a number of very lengthy demonstration experiments have been cut out.

Mathematics in chemistry

Almost all schools pursuing the scheme reported that their pupils experienced difficulty in using mathematics in their work. There is no reason to believe that this difficulty is any more pronounced with pupils following a Nuffield course than any other course, but it was clear that any way in which mathematics and chemistry could be brought together more closely would be worth while. For this reason, *Teachers' guide II* incorporates a number of notes on the use of mathematics in chemistry.

Options

The idea of optional studies to round off the course has been met with widespread approval, although some schools have been somewhat pushed to complete the second option. Some of the options have been very much more popular than others – *Water* and *Periodicity* have been taken by the largest number of schools.

Some schools have encouraged their pupils to express some choice in which options they would like to study – a few schools reset their classes into groups which study different options according to the choice made by individual pupils. A limiting factor in getting pupils to choose their options has been the existence of background material scattered in a number of background books.

Social consequences of the work of the chemist

The attitude of science teachers is not unanimous at the present time on the desirability of trying to make pupils aware of the social relevance of the subject, as part of the teaching of chemistry. It is clear that an increasing number of teachers would like to see more suggestions for topics including social relevance incorporated within the Sample Schemes, and accordingly there is new emphasis on the social side of the subject, particularly in *Teachers' guide II* and in the *Handbook for pupils*.

The need to instil confidence

There was evidence of a 'lack of security' among some teachers and pupils, which often emerged in remarks such as 'We don't quite know what is required of us'. It would be impossible to say whether this was a more prevalent view among teachers following the Nuffield scheme than among those following other courses, but it showed the importance of looking at the materials to see whether the purposes of pursuing a particular piece of work emerge as clearly as possible. This is one reason why we have introduced *objectives* in the *Teachers' guide* although we hope that these will be looked upon more as suggestions than as directives.

The lack of a single pupils' book has been an understandable source of worry to some pupils, particularly those who are at any sort of disadvantage, such as sometimes result from moving from a school pursuing another type of chemistry course, or from illness over an extended period. It is also hoped that the *Handbook for pupils* (see Chapter 6) will be a great help in giving the pupil confidence.

Other studies

The present revision has been based mainly on a study of the experiences of teachers and pupils using the schemes, but several other studies have been made which have had a bearing on the revision, of which two will be mentioned here.

1. The conceptual demands of Stage IA and Stage II have been analysed* and the results of this study have further emphasized the high level of formal thinking which is required of pupils at the beginning of Stage II.
2. The performance of pupils in the examination has also been studied. This has given information on how pupils who were entered for the Nuffield examination performed in each of the various Topics of the course. In general, the results of this study have shown that the great majority of the work in the Sample Scheme has been within the capacity of pupils who entered for the examination.

*Ingle, R. B. and Shayer, M. (1971) 'Conceptual demands in Nuffield O-level Chemistry'. *Education in chemistry*, **8**, 182.

Multiplicity of curriculum materials

The very large number of materials – for example, over fifty books including the Background Books, and over thirty film loops – was one of the merits of the first edition, providing exceptional flexibility of choice both in subjects and in use. This very multiplicity, however, may have limited the impact of the scheme as a whole in certain areas. For this reason we have looked very carefully at the series of books with the aim of reducing the number of titles to keep down costs and to make the scheme more manageable. We have also given the new books titles that clearly indicate their function. This reduction in the number of titles particularly affects the books for pupils.

Pupils' reading material

The originators of the first edition took the remarkable and far-sighted step of *not* producing a single textbook. Their intentions were made clear in the *Introduction and guide*:

'We have in fact tried to analyse the purpose of a textbook and break it down according to its functions. This analysis has given us the *Book of Data, Laboratory Investigations*, and the Background Books. From the *Book of Data* the pupil can see if his ideas fit observed facts. From *Laboratory Investigations* he builds up his own part of the textbook. From the Background Books he forms his own library.'

It was obviously very desirable to see to what extent these intentions had been realized in schools. Taking in turn each of these publications:

Book of Data. Although there was evidence that pupils found parts of it interesting, feedback was almost unanimous that it was 'too complicated', and that it contained more information than was required at this level. A careful study has been made of the difficulties encountered by pupils in learning to use a book of data. As a result of this most of the data is now presented in the form of tables each of which contain a strictly limited amount of information, and these form the last part of the new *Handbook for pupils*.

Laboratory Investigations. These sheets have been widely used, either as they stand or after adaptation. They were written for use in streamed classes, the situation common to almost all schools in the early and mid 1960s, and they did not seek to give full instructions for pupils. Not surprisingly, they have been less well received in schools in which there is a wider range of pupil abilities in a single class, and teachers in this sort of situation usually had to make extensive modifications to them.

We have therefore in the light of this development revised the sheets and renamed them *Experiment sheets*. We offer a revised sample which is rather more detailed and explicit, in the belief that those teachers who do still feel the need of some published sheets, will probably find these more useful than the original ones. Many teachers have now written their own sheets for the entire five-year course. Others prefer to teach without using them at all, on the grounds that they are unnecessary or tend to stereotype the lesson by encouraging excessive preoccupation with following instructions. There is, certainly, some evidence that the use of worksheets is counter-productive to the open-minded spirit of investigation by pupils desired by many teachers. Yet a majority of teachers at the present time find some form of worksheet a necessity, particularly when they are working with a wide range of ability of pupils in the same class.

Teachers who think of producing their own worksheets may care to compare the original edition of *Laboratory Investigations* with the revised *Experiment sheets*. The Nuffield Secondary Science *Teachers' guide**, pages 63–77, discusses the production of different styles of sheets for different situations and purposes, and the *Scottish Integrated Science Project*† sheets have used a variety of styles from which much can be learnt.

Background Books. Although these books have circulated widely and have been much used, the original hope that pupils would build up their own library of these books has seldom been fulfilled. The relatively high cost of a set of these books, compared with that of a single pupils' book, has meant that in most schools one set of books has had to be shared between several classes. The business of continually giving them out, and more especially of collecting them, has sometimes proved a considerable disadvantage – in some cases so much so that they are only read in class time and not taken home.

The conclusion was reluctantly reached that the particular way in which the functions of a textbook were broken down into Laboratory Investigations, a Book of Data, and Background Books, had not been completely successful. For this reason, the pupils' materials have been restructured in an attempt to achieve the original aims of the project, but by slightly different means. The way in which this has been done in providing *Study sheets* for Stage I and the *Handbook for pupils* and *Chemists in the world* for Stage II is outlined in the following chapter which describes the books and materials now available in the new edition.

*Nuffield Secondary Science Project (1971) *Teachers' guide*. Longman.
†Scottish Integrated Science Project (1973) Science for the seventies, *Science Worksheets*. Heinemann Educational.

Chapter 6

Publications and teaching aids

To achieve a new approach to education through chemistry, the Nuffield Chemistry Project in 1966 produced a series of publications in line with the proposals discussed so far in Part 1. The basic aim of the project remains the same, but in this second edition many alterations have been made to the range of publications now familiar to many teachers. A description of the functions of the books in the new edition will bring out further the ways described in the last chapter in which the revision has affected the Sample Scheme as a whole.

Books for the teacher

Five books are now available which offer advice to teachers on planning their teaching schemes, and on the teaching resources available to them. These books, to which many teachers have contributed, have two main purposes:
1. They enlarge upon the general aims of the scheme (as described briefly in Part 1 of this book), with suggestions about topics of study and how to handle them and about the scope and content of possible courses arising out of the scheme.
2. They help teachers to adapt the scheme to the particular needs of their pupils, through suggestions about teaching approaches and the organization of lessons, and about the availability of teaching aids.

Titles and details of these books are as follows:

Teachers' guide I
Teachers' guide II
These two volumes are self-contained books, Volume I dealing with Stage I (years 1 and 2) and Volume II dealing with Stage II (years 3, 4, and part of 5) of the five-year course leading to the O-level examination. They take the place of the *Introduction and guide* and *The Sample Scheme Stages I and II: The Basic Course*. Full details of these new books are given in the Editor's Preface to this book.

Teachers' guide III
This book contains the revised proposals for the options in the last year of the five-year course and is in parallel with the pupils' options.

Handbook for teachers
This book has not been revised. It contains essays covering the intellectual and theoretical grounds on which the scheme is based together with much information and advice. Its main objects are:
1. To show how the major ideas of present-day chemistry, those concerned with structure, rates, equilibria, and energy changes, are integrated in the Sample Scheme.

2. To examine the possible consequences of introducing these topics to work in the sixth form. Its contents therefore in some of the Topics deliberately go beyond that needed for the 11–16 age group. It is not intended solely as a guide to the Sample Scheme; those wishing to devise other teaching programmes should find much that is helpful in its pages. No attempt is made to avoid controversial issues, but these are discussed in a constructive manner in the hope that a contribution can be made to future progress in the teaching of chemistry.

Collected experiments
This book has also not been revised but it is still of considerable value. It is for the use of teachers and has two purposes.
1. To provide alternative experiments in case those given in the *Teachers' guides* are not suited to the particular conditions under which the teacher is working.
2. To provide some experimental materials for teachers who want to use the Nuffield approach, but who want to teach subject matter other than that given in the *Teachers' guides*.

The experiments are arranged in this book by 'themes', that is, experiments on related topics are grouped together.

Books for the pupil

As a result of the inquiries on which this revision is based the books for pupils have undergone major changes in content and format, changes made on the one hand to make them accord with requests and suggestions from schools and on the other to make their costs as reasonable as possible. We have still kept, however, to the original aim of not basing the course upon a single textbook which might impose a rigid scheme of study. The editors of the first edition were quoted in the last chapter, saying that they had tried to analyse the purpose of a textbook and break it down according to its functions, and they provided titles to fit those functions. Our subsequent evaluation has shown new ways of keeping this open approach but interpreting it according to the needs of schools today. As a result of this the *Book of Data* and many of the Background Books will have much of their content incorporated in the *Handbook for pupils* and *Chemists in the world*. These new publications contain the relevant material from their predecessors but deal with subjects in more depth. The most notable new work is the *Handbook for pupils* which is described below.

The titles of the books for pupils are:

Stage I { *Experiment sheets I*, *Study sheets*, *Growing crystals* (Background book) }

Stage II { *Experiment sheets II*, *Handbook for pupils*, *Chemists in the world* }

Stage III Option booklets

Experiment sheets I and II
These titles replace the *Laboratory Investigations I* and *II* of the first edition. The format of these titles has been changed from that of the first edition because they were inconvenient to distribute in class, and also expensive. They are now made available as pamphlets in A5 format. Each page is punched. With this new format the sheets can be used in two ways: first, individual pages can easily be cut out of the pamphlets and given out at the appropriate time. Used in this way the sheets can either be (1) inserted in ring binders, or (2) gummed into notebooks, since the page size has been reduced to fit standard sizes of notebooks. Alternatively, the pamphlets can be kept complete.

Experiment sheets I contains the sheets for both IA and IB. This will help the many teachers who, as the evaluation has shown, make their own routes using material from both alternatives. *Experiment sheets II* similarly contains the instructions for IIA and the new alternative IIB.

The content of *Experiment sheets* has been thoroughly rewritten in the light of comments from schools. As a further help to the teacher each sheet is reproduced in the *Teachers' guides*.

Study sheets
The twelve Study sheets replace the six Background Books originally designed for Stage I. They have been devised to give a more balanced coverage of background reading for Stage I. Suggestions for when they could be used are given in Part 2 of this *Teachers' guide*.

They have also been devised as a means of fulfilling one of the original aims of the course described above, that the child should build up his own textbook. It was originally hoped that the Background Books could be given away to the children so they could keep them and continue to profit from them. Their format proved too expensive for this to happen. The Study sheet format allows a considerable reduction in cost while providing a fuller range of reading. The response of a dozen schools to trial versions of two Study sheets indicated that there was a general need for cheap and, if possible, disposable background reading of this kind.

They are available as punched folded sheets in packs containing ten of each Study sheet. As with the new *Experiment sheets* their format allows flexibility of use. If given out to individual members of the class they can be inserted in ring binders or gummed into notebooks. Otherwise they can be inserted in cardboard folders and given out as class sets.

The titles of the Study sheets are as follows. Those marked * are printed in full colour.

Analysis
*Where chemicals come from**
Heating things
Burning and Lavoisier

Fresh air?
*The chemical elements**
Competitions
*Water**
Chemistry and electricity
*Chemicals and rocks**
The halogens
The words chemists use.

Short descriptions of each Study sheet accompany the directions in Part 2 of the *Teachers' guide* on when to use them.

Handbook for pupils
This book for Stage II has been written in response to requests from numerous schools for a more solid backing for pupils. It consists of three parts, the first containing chapters on topics of chemistry, ranging from periodicity to structure, to which the teacher can refer a child needing an exposition of a particular chemical idea. The second part contains reading material on the chemical industry, the world food problem, and the social implications of chemistry. The third part contains the basic data necessary for the course, together with a glossary and answers.

The greatest effort has been expended to make this book as attractive as possible while at the same time ensuring that the diagrams and illustrations convey information simply and clearly. The chapter on structure and the three chapters of Part II are illustrated in full colour. A second colour is used in the rest of the book. Trials were conducted on the chapters on structure and on social implications.

Chemists in the world
This book consists of the material from the Background Books for Stage II not incorporated in the *Handbook for pupils*. The material has been thoroughly edited and brought up to date where necessary. It has two parts. The first covers the history of chemistry from the development of the atomic theory by Dalton to modern times. The second deals with the industrial applications of chemical discoveries concentrating mainly on the uses of coal and petroleum in providing energy and materials.

Pupils' options
For details of the new format of the pupils' options and the background reading for them, consult the publishers' brochure.

Visual aids

Film loops
Stage I
1–1 Salt production
1–2 Chlorophyll extraction
1–3 Whisky distillation
1–4 Oil prospecting
1–5 Petroleum fractionation
1–6 Liquid air fractionation

1–7 Gold mining
1–8 Iron extraction
1–9 Copper refining
1–10 Limestone
1–11 Fluorine manufacture
1–12 Uses of fluorine compounds
1–13 Chlorine manufacture
1–14 Chlorine – uses
1–15 Bromine manufacture
1–16 Bromine – uses
1–17 Iodine manufacture
1–18 Iodine – uses

Stages II and III
For details of these film loops, consult the publishers' brochure.

Overhead projection originals
These are for Stages II and III. They are a set of printed originals which can be used for making transparencies for use on overhead projectors, for duplicating and distributing to the class, or as small-scale wallcharts.

Slide series
Details of these and of their availability will appear in the publishers' brochure.

Part 2 Alternative A

Topic A1

Getting pure substances from the world around us

Purposes of the Topic

The main purpose of this Topic is to show the pupils what a chemist means by a pure substance.

The pupil starting this course will have little idea of what is meant by a pure substance, beyond the expression 'guaranteed pure' seen in the shops. During this Topic he should learn:

1. The meanings of the terms 'pure substance' and 'mixture'. His understanding of the concept of a pure substance will grow throughout Stage I and well beyond.
2. Something about some simple techniques for separating substances – notably paper chromatography, distillation, filtration, evaporation, and crystallization.
3. The uses of words by experience rather than by formal definition. Lists of new words are given at the end of each section.
4. How the experiments in the laboratory have their counterparts in the practical world of industry, and the importance of chemistry in everyday life.

Contents

Timing

Each major experiment will require at least one double period. With a form of average ability, eight or nine double periods will be needed to do the Topic justice.

Introduction to the Topic

After some preliminary work with a Bunsen burner, the pupils try to separate 'pure' substances from familiar materials. They separate pure salt from rock salt, water from ink, and chlorophyll from grass. At this point use is made of the fact that plant pigments have been introduced to depart from the main theme and study the effect of some everyday substances on indicators. In this way acidity–alkalinity is introduced in purely operational terms. The pupils learn to recognize acidity or alkalinity with the aid of Full-range Indicator and to identify a particular degree of acidity in terms of the colour of the indicator and a number, the pH number, which is associated with this colour.

The distillation of crude oil introduces the subject of fractionation as a means of separating liquids and of determining whether a substance is a single ('pure') substance or a mixture. This return to the main theme of the Topic is followed by the extraction of lead from lead carbonate ('cerrusite'). The pupils find that the lead is not present *as lead* in the 'ore', but is present in a quite different form. They have to use a special method involving another substance, carbon, in order to be able to extract the lead. In this way they are introduced to compounds.

Alternative approach

A different treatment of the same theme is given in Alternative B, Topic B1.

Subsequent development

Most of the experimental techniques learned in this Topic are basic to the whole of the pupils' later experimental work during the next two years, and time spent here is thus well worth while. These techniques include filtration, evaporation, crystallization, distillation (simple and fractional), paper chromatography, heating on asbestos paper, and electrolysis. In particular, heating on asbestos paper is widely used in Topic A6, and electrolysis is studied further in Topic A8.

Other knowledge gained in this Topic is used as follows. The pH scale and the idea of acidity–alkalinity are needed in Topic A4 during the investigation of the products of combustion. The knowledge of what is meant by a pure substance is basic to the rest of the course, and the differences between mixtures and compounds are brought out throughout the rest of the two years' work.

Further references

for the teacher

Other experiments illustrating the same theme can be found in *Collected experiments*, Chapter 2 'Isolation and recognition of a single substance'.

A general discussion of the investigational method, on which most work in the Nuffield Chemistry scheme is based, will be found in Chapter 1 of the *Handbook for teachers.* A general discussion on approaches and attitudes to starting a chemistry course is given in Chapter 3 of the same book.

Physics

Teachers whose classes are studying the Nuffield Physics course should read the Physics *Teachers' guide I.*

Supplementary materials

Film loops

1–1 'Salt production'
1–2 'Chlorophyll extraction'
1–3 'Whisky distillation'
1–4 'Oil prospecting'
1–5 'Petroleum fractionation'
1–7 'Gold mining'
1–9 'Copper refining'

Films

'Refining'
'Oil'
'North Sea strike'
'Treasure trove'
BBC programme 'Coal age'

See Appendix 4, page 242, for brief descriptions and further details of films and BBC TV programmes

Reading material

for the pupil

Study sheets:

Analysis. In this, pupils are helped, by pictures and words, to understand what analysis means to a chemist. They are also shown how analysts can help us by checking samples of food and air, for example, and how forensic scientists can provide vital clues in criminal investigations.

Where chemicals come from describes how the chemicals the pupils use, and many others besides, are obtained from air, sea, rocks, and plants. It covers a variety of substances and its aim is to widen the children's appreciation of the extraordinary wealth of Nature.

The words chemists use contains a glossary of expressions used in Stage I, a short account of the derivation of various words, and a table of dates and discoverers of some elements. It can be used at any part of Stage I but many teachers may like to use it at the beginning as an introduction to the new and, perhaps, strange words that pupils will meet.

Background book:
Growing crystals

Starting the course – the first lessons

There are many ways of starting a chemistry course. One way to capture the pupils' interest and curiosity in the subject is to have on the bench a large number of everyday substances: detergent, a brick, a piece of marble, some paint, a bottle of aspirins, and indeed anything that is relevant and readily available. You can then explain to the class that chemistry is concerned with all these things – finding out about the common materials around us, and then shaping them to our own needs. More is said later about this approach.

During the introductory lesson or lessons the pupils will, of course, have to be shown the laboratory and possibly the science library. A certain amount of administrative work will no doubt also be done so that this introduction may take either one or two double periods.

Before starting the course proper it is worth spending a lesson, or part of a lesson, in discussing and learning to use the Bunsen burner. The pupils will have to use the burner in the first experiment that they do and it is more convenient to deal with it in advance than to have to stop in the middle of an experiment to do so.

The Bunsen burner – a suggested approach

Start by telling the class of the importance of the Bunsen burner in simple laboratory chemistry. It is remarkable that this piece of apparatus, which Bunsen invented for his new laboratory in Heidelberg in 1865, remains virtually unaltered today. Tell the pupils to close the air-hole, turn the gas full on and light it. Now they can all try slowly opening the hole. What do they notice? How many things change when the air-hole is opened? The luminosity of the flame (the question 'why?' can be answered later); the length of the flame; its temperature (ask them if it seems to get hotter); the noise it makes (ask them why). Now get the pupils to open the air-hole fully and gradually turn the gas down until the flame strikes back*. Some will notice that the flame is still burning at the bottom of the burner. Tell them to turn the gas off, close the air-hole (with care as the barrel may be very hot), and relight the burner. They may now examine the cause of the luminosity of the flame by closing the air-hole again and holding a piece of broken porcelain (from, say, a crucible lid) in the flame with a pair of tongs. Ask them to note what happens. Now get them to hold the blackened porcelain in a roaring flame. They will see that the black deposit disappears.

**This may not happen with North Sea gas.*

Another way to open the lesson on the Bunsen burner is to arrange that half the Bunsen air-holes are open and half closed before the lesson starts. Let the pupils light their Bunsen burners and then point out that half of them have a luminous flame and half have a non-luminous flame. Ask them why this is so.

Let the pupils have a closer look at the flame itself. The characteristics of the different parts of the flame are nicely shown by using a piece of asbestos paper. Hold the paper flat in the flame (air-hole open) and see which part of the flame is hot enough to make the asbestos red-hot. Now lift the paper up through the flame and watch the rings decrease in size. Ask them which they think is the hottest part of the flame.

Let the pupils practise setting the burner with the gas *half* on and with the air-hole *half* open. This is the kind of flame which they will usually require and is also the flame we shall mean when we say 'heat with a Bunsen flame'.

A1.1 How can we get pure salt from rock salt?

For their first example of separating pure materials from naturally occurring sources, the pupils are led by the teacher to devise a method for making pure table salt from rock salt. They are introduced to the techniques of solution, filtration, evaporation, and crystallization.

A suggested approach

Objectives for pupils*

1. Skill in the techniques of filtration, evaporation, and crystallization
2. Awareness of what the chemist means by a 'pure' substance
3. Knowledge of the process of salt production
4. Understanding of the meaning of the words: dissolve, solvent, solution, filter, filtrate, evaporate, crystal, crystallize

The teacher will need

Sample of rock salt

Film loop projector

Film loop 1-1 'Salt production' (with notes)

The first lessons have introduced the pupils to the idea that chemists are concerned with the materials or 'stuffs' of which the world is made. Tell the pupils now that before they can find out about these materials they must learn to distinguish between a mixture of several substances and a single substance. To begin with, the pupils have no criteria for doing this and they do not know what is meant by a single substance. They cannot tell whether wood, for instance, is a mixture of different materials or one single stuff. In fact most children of this age think of wood as a single substance.

The pupils can, however, distinguish between 'clean' and 'dirty' substances, and if we show them some 'rock salt' and say that table salt is produced from it, they may appreciate that rock salt is a 'dirty' form of table salt. In other words it contains things other than the pure salt that the salt manufacturer wants to get out of it. If they look at rock salt under a hand lens they will be able to see that there are 'sand-like' particles mixed up with crystals and that some of the crystals are colourless and look 'clean'. Ask them 'How can we get pure table salt from the rock salt?' They may suggest washing away the sand. Let them try this with a small piece in a test-tube. They will find that it is the salt which 'disappears', and that the sand remains behind.

*For the sense in which this word is used, see Part 1, page 13.

Most of them will be familiar with the idea of dissolving from such processes as the dissolving of sugar in tea and salt in soup, and this is a convenient time to introduce the word 'dissolve' to those who are not familiar with it. Pupils can be told that dissolving is a useful and important process for chemists.

The teacher's function as far as possible should be to help the lesson along by asking questions. 'Is the salt dissolving quickly or slowly?' 'Would smaller pieces dissolve more quickly?' – here the pestle and mortar can be demonstrated. 'Dissolving rock salt in water will take out the salt, and crushing the lumps first with a pestle and mortar will speed the process up; but how is the sandy stuff to be separated from the solution?' The pupils will probably suggest 'straining', which introduces filtration. Filter paper does indeed act as a strainer, and the teacher should demonstrate how to fold it, and place it in a funnel. 'What has happened to the salt?' – of course it is in the water. 'How can we get it back?' Someone may suggest 'driving off' the water. Pupils can try this on a small scale on microscope slides, or the teacher can demonstrate it with a drop of salt solution on a slide or spatula. The pupils should look at the residue through a hand lens.

The words solution and solvent should also be introduced at this point. Pupils should not be introduced to too many new words at one time, but they should be able to assimilate half a dozen or so in each lesson. They understand the words best by using them in the course of experiments. They should not be expected to memorize dictated definitions.

Getting rid of the water entails evaporation, and how this is done on a larger scale should be shown. A steam bath is used for the later stages and its use as a means of 'driving off' water more gently should be explained and demonstrated.

Once the experiment has been planned the pupils can start work.

Experiment A1.1

Production of pure salt from rock salt

Apparatus

Each pupil (or pair) will need:

Experiment sheet 1

Beaker, 100 cm^3

Pestle and mortar

Glass stirring rod

Filter funnel and paper

Funnel stand

Hand lens

Tripod stand and gauze

Bunsen burner and asbestos square

(continued)

Procedure

This is described in *Experiment sheet* 1 from the pupils' book, *Experiment sheets I*. To aid the teacher this sheet is also reproduced here. (All the *Experiment sheets* are reproduced at the relevant points in this book, and are distinguished from the rest of the text by having a line round them.)

Heating the filtered solution in the evaporating basin is best done by placing the basin on a gauze on a tripod stand over a Bunsen burner until most of the liquid has gone. The final stages of the evaporation however should be done more carefully, as the solid is likely to jump out. It is most convenient for the teacher to have a multi-position electrically-heated steam bath on which pupils can place their basins for final evaporation. Alternatively, each pupil can use a small beaker half full of water placed on a gauze on a tripod stand and heated by a Bunsen burner.

Access to an electrically-heated water bath containing at least 18 positions (or beaker, small enough for the evaporating basin to rest on top)

Sample of rock salt.
Note. The salt should contain obvious solid impurities. Salt used for removing ice from roads is ideal and may be obtainable from the Local Authority Highways Department.

See the suggestion for building up a pupil's record of experiments in Appendix 2, page 234.

Experiment sheet 1
Rock salt contains other substances in addition to salt, which make it impure. In this experiment you are going to try to get pure salt from rock salt.

To a 100 cm^3 beaker add enough rock salt crystals to cover the bottom of the beaker. Pour the rock salt into a mortar and grind it into powder using a pestle. Pour the powdered salt back into the beaker and add 30 to 35 cm^3 of water. Place the beaker on a wire gauze on a tripod stand and heat it with a medium-size Bunsen burner flame, stirring the liquid with a glass rod. When no more salt seems to dissolve, filter the mixture while still hot (use a cloth to hold the beaker) into an evaporating basin.

Heat the solution in the basin until all the water has been driven off. (Your teacher will tell you how to do this.)

When the dry salt is cool put a few crystals of it on to a piece of paper. Look at them through a hand lens or a low-power microscope. What shape are the crystals?

Draw one of the crystals in the space below.

While the pupils are evaporating the salt solution the film loop 1–1 'Salt production' can be shown. This is the first example in the course of the use of energy to bring about a desired change. Both during the experiment and when showing the film loop this should be stressed; see the *Handbook for teachers*, Chapter 15. The teacher should be familiar with the film before showing it to the class. Full details of the film loop and background information about salt production are given in the notes supplied. This is an important film loop as it shows the pupils at an early stage that the chemistry they are studying in the laboratory has applications in the world outside.

After purifying the salt the pupils are asked to examine the crystals under a hand lens. They should note the shape and regularity of the crystals. Introduce the words 'crystal', 'crystallize', and 'evaporation' in the discussion, and see that pupils realize that driving off the water does not remove anything but the solvent.

New words

Dissolve	Filtrate
Solvent	Evaporate
Solution	Crystal
Filter	Crystallize

Suggestions for homework

1. How would you show that soil contains some substances that are soluble in water?
2. How could you find out if your tap water at home is pure?
3. Read the Background book, *Growing crystals*.
4. A number of words are used in chemistry which are not in common use elsewhere, and a number of everyday words have special meanings in chemistry. Start to collect new words in a 'Chemistry word book' in alphabetical order for easy reference.

Summary

Pupils should now be able to use the techniques of filtration, evaporation, and crystallization. They should know how salt is made industrially from rock salt; understand the meaning of the words dissolve, solvent, solution, filter, filtrate, evaporate, crystal, crystallize; and they should have some awareness of what is meant by a pure substance (namely, a single substance).

A1.2
Is ink a single substance?

In this section the pupils continue the study of what is meant by a pure substance by seeing if they can separate ink into several substances.

A suggested approach

Objectives for pupils

1. Skill in the techniques of simple distillation and paper chromatography
2. Understanding what the chemist means by a 'pure' substance
3. Knowledge of the process of distillation as exemplified by the production of whisky
4. Knowledge of how to test a liquid to see if it is water
5. Understanding of the meaning of the words: distil, distillation, condense, condenser, thermometer, chromatography, chromatogram

The teacher will need

Film loop projector

Film loop 1–3 'Whisky distillation' (with notes)

Start by asking the pupils what the last experiment told them about rock salt. 'Is rock salt a single substance?' It is a mixture of at least least two substances and probably many more. Now ask how they would try to find out whether ink is a single substance or not. Someone may suggest filtering. Try this or let the class try it. Most inks come through the paper unchanged.

Now suggest boiling the ink to see what happens. Ask the pupils if they would expect a black or blue vapour to come off. In the fume cupboard pour a little bromine into a test-tube and warm it gently. The vapour will be seen to be the same colour as the liquid. Now return to the ink, and boil a little in a test-tube. Pupils will see that the vapour is not coloured. If a little of the vapour is condensed on a cold surface (an evaporating dish or a test-tube full of cold water), they will see that a colourless liquid is formed. This suggests that ink can be separated into at least two components by boiling it and then condensing the vapour. Ask the pupils how they would design an apparatus to do this on a larger scale. With some help and discussion, the apparatus shown below may be devised and used for a simple pupil experiment. During the discussion introduce the words: 'distillation', 'condense', and 'condenser'.

Experiment A1.2a

Boiling ink

Apparatus

Each pupil (or pair) will need:

Experiment sheet 2

Conical flask, 100 cm^3

Cork or bung with one hole, to fit the flask

Length of glass tubing bent as in the diagram

Test-tube, 100 × 16 mm

Measuring cylinder, 25 cm^3

Stand and 2 clamps

Tripod stand and gauze

Bunsen burner and asbestos square

Supply of ink (Stephen's black, or black Quink inks are suitable. Some inks froth excessively when boiled, and so a trial distillation before the lesson is advised.)

Procedure

This is described in *Experiment sheet* 2 reproduced below.

Experiment sheet 2

In the last experiment, water was boiled away from a salt solution and salt crystals left behind. You will now find out what happens when another liquid, ink, is boiled. You are going to try to find out whether ink is a single substance or not.

A suitable apparatus is shown in the diagram but other arrangements can be used.

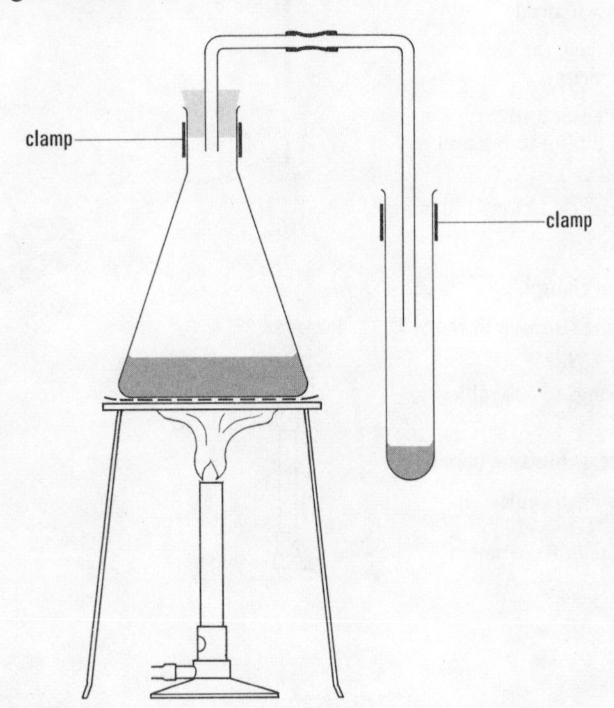

Use about 10 cm^3 of ink. Place it in the conical flask, and insert the cork carrying the glass tubing firmly in the mouth of the flask. Place the flask on the gauze on the tripod stand and clamp it in position as shown.

Heat the flask with a small Bunsen flame until the ink boils gently. If the ink froths up the flask stop heating until it settles down again.

What do you notice at the end of the glass tube which is in the test-tube?

Continue heating for a minute or two.
What is happening in the test-tube now?

Stop heating before the flask becomes dry.
Is ink a single substance or a mixture of substances?

A large yield of water is not needed. When some has been collected, discuss the results with the class. What was wrong with the apparatus? The ink splashed over and a lot of the water was lost as steam. A more efficient condenser is needed. Show the class a Liebig condenser. Like the Bunsen burner, this simple but efficient laboratory condenser has stood the test of time. Now demonstrate the distillation of ink using a distillation flask and a Liebig condenser, as described below.

Experiment A1.2b

The distillation of ink

This experiment should be done by the teacher.

Apparatus

The teacher will need:

Distillation flask (at least 100 cm^3 capacity)

Liebig condenser and connection tubing to tap and waste

Thermometer, −10 to +110 °C

Beaker, 100 cm^3

2 stands and clamps

Tripod, gauze, Bunsen burner, and asbestos square

Corks or bungs to assemble apparatus*

Filter paper or blotting paper

Anti-bumping granules

Ink (see note in Experiment A1.2a)

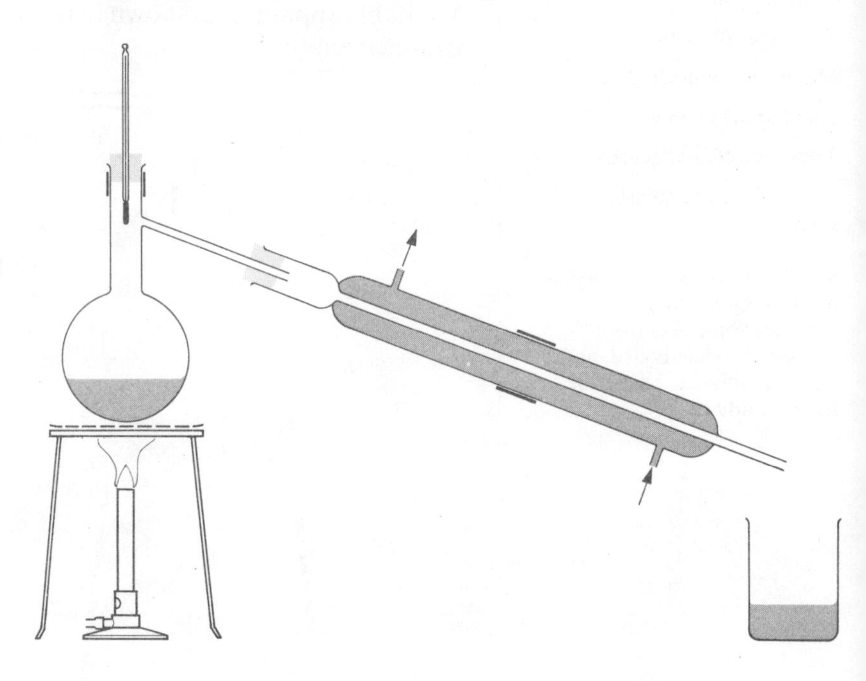

Procedure

The distillation apparatus should be set up as shown in the diagram. Important points to note are:

1. The thermometer bulb should be just below the side-arm of the distillation flask to record the temperature of the vapour actually being collected.
2. The cooling water should enter through the lower inlet and leave by the upper inlet so as to keep the condenser jacket full of cooling water.
3. A few anti-bumping granules should be placed in the distillation flask to ensure even boiling. (There is an opening here for a discussion on the reason why.)

*Corks or bungs will not be needed if apparatus fitted with interchangeable ground-glass joints (for example, Quickfit apparatus) is used. Although more expensive, this type of apparatus is much more convenient, and is recommended for the teacher's own use.

Collect sufficient water for all the pupils to see, but do not boil the contents of the flask dry, or it may crack.

Finally, mix the water with the contents of the flask to show that 'ink' is formed again.

It has been assumed that the liquid collected is water, and most pupils will assume this also. But how do they know, and how can they tell, that it is in fact water? Suggest that one way to show that it is water is to measure its boiling point to see if it boils at 100 °C. Many pupils of this age still have difficulty in believing that there are colourless liquids which are not, or do not contain, water. It may be worth demonstrating a simple measurement of the boiling point of a liquid other than water at this stage. Details of such an experiment are given below.

Experiment A1.2c

How to measure boiling points

This experiment should be done by the teacher.

Apparatus

The teacher will need:

Beaker, 250 cm³

Bunsen burner and asbestos square

Tripod and gauze

Hard-glass test-tube, 150 × 25 mm

Thermometer, −10 to +110 °C

Stand and 2 clamps

Anti-bumping granules

Ethanol or acetone

Procedure

Place a 250 cm³ beaker half full of water on a tripod and gauze, over a Bunsen burner, and by means of a stand and two clamps support a 150 × 25 mm test-tube containing not more than 10 cm³ of ethanol in the beaker, and a thermometer in the test-tube so that its bulb is held just above the level of the ethanol. Put one or two anti-bumping granules in the ethanol. Heat the water in the beaker to about 80 °C and when the ethanol begins to boil note the steady reading on the thermometer. *Ethanol should not be heated directly by the Bunsen burner because of the risk of fire*. As an alternative, acetone (which is also flammable) may be used in place of ethanol. Boiling points are: ethanol 78 °C, acetone 56 °C.

Finally, find the boiling point of the 'water' collected in Experiment A1.2b. Eventually, the test-tube will need to be heated directly by means of the Bunsen burner for this measurement.

The industrial importance of distillation should be mentioned, and the production of whisky can be given as an example. Film loop 1–3 'Whisky distillation' will be of help here, and the notes accompanying the loop should be read carefully before it is shown.

Another technique for separating mixtures can now be introduced by asking the pupils if they have ever noticed what happens if a drop of ink falls onto blotting paper. Let them try this with ink from their own pens on blotting paper or filter paper. They will see that the colours of the various dyes (black ink shows this particularly well) separate as they spread out. It may be a surprise to many of the pupils that some black ink (for example, Quink Permanent Black) actually contains orange and blue dyes. A drop or two of water placed on the ink blot will spread the dyes out. This simple experiment shows that there is more to the question 'Is ink a single substance?' than may have appeared to the pupils at first glance.

Explain to the pupils how this separation can be improved, and introduce the words 'chromatography' (for the process) and 'chromatogram' (for the paper with the separated colours on it). They can then do the next experiment.

Experiment A1.2d

The chromatography of ink

Apparatus

Each pupil (or pair) will need:

Experiment sheet 3

Filter paper

Beaker, 250 cm^3 or evaporating basin

Beaker, 100 cm^3

Teat pipette

Supply of Quink Permanent Black Ink

Procedure

This is described in *Experiment sheet* 3 which is reproduced below.

Experiment sheet 3

From *Experiment* 2 'Boiling ink', you have seen that ink contains at least two substances, water and colouring matter. Do you think the colouring matter is a single substance or a mixture? This experiment will help you to find out.

Rest a filter paper on the top of a beaker or evaporating basin. Put *one small* drop of ink on to the centre of the filter paper. Add separate drops of water to the ink spot, allowing each one to spread out before adding the next.
What happens?
Is the colouring matter in the ink a single substance or a mixture?

When the water reaches the edge of the filter paper stop the addition of drops and allow the paper to dry.

Draw a diagram showing the appearance of the chromatogram obtained.

The chromatogram obtained from this experiment will contain blue and yellow bands. The pupils can achieve a complete separation by cutting the filter paper with scissors between the bands. The teacher should arrange to have two beakers on the front bench, and collect the blue portions of filter paper in one, and the yellow in the other. A small quantity of water put in each beaker will extract the colour and complete separation will have been achieved. (The extraction is improved by warming with a little acetone.)

New words

Distil
Distillation
Condense
Condenser
Thermometer
Chromatography
Chromatogram

Suggestions for homework

1. Design a condenser for yourself. In what ways do you think it is better than a Liebig condenser?
2. Make a list of all the processes you can think of which make use of distillation. Illustrate your list with cuttings from magazines.
3. Your friend asks you for some distilled water for his car battery. Explain to him how he can make his own from tap water using ordinary kitchen equipment.

4. Suggest a way of making drinking water out of sea water on a large scale in a tropical country. (There are countries where this problem has had to be solved.)

Summary

Pupils should now be able to use the techniques of simple distillation and paper chromatography. They should know the part distillation has to play in the production of whisky, and how boiling points can be used to identify substances. They should understand the meaning of the words distil, distillation, condense, condenser, thermometer, chromatography, chromatogram.

A1.3
What coloured substances can we get from grass?

In this section the pupils follow the search for pure substances by trying first to extract the 'green stuff' from grass and then seeing if this 'green stuff' itself is pure.

A suggested approach

Objectives for pupils

1. Ability to select the appropriate experimental technique
2. Skill in the technique of paper chromatography
3. Use of the chemist's idea of 'pure' substance
4. Knowledge of the industrial process of chlorophyll extraction

The teacher will need

Film loop projector

Film loop 1–2 'Chlorophyll extraction' (with notes)

Is grass* green because it is made entirely of one green substance or is there a green stuff which colours it? Some pupils may recall getting green stains on their clothes, and this may make them think that the green colour comes out of the grass. In any case, ask the pupils how they could find the answer to this question: 'Is it possible for you to dissolve the green from the grass in the same way that you got the salt out of rock salt?' They will soon come to the conclusion that, if this were the case, the grass would have its colour washed out by the rain. But what if the grass is crushed? Get the class to try crushing a handful of cut grass with a little water in a pestle and mortar. They may find that after some time the water becomes green. If it does, get them to filter the 'green water'. The green colour remains on the filter paper. This experiment has (in one sense) failed. Ask the class for more ideas. If the green stuff (assuming there is such a material) does not dissolve in water, will it dissolve in anything else? If water does not get the grass stain out of clothes, what does? What about dry cleaning? There are solvents other than water. Show them two of these: acetone and ethanol. (*Caution:* both these solvents are flammable.) Now get the pupils to try crushing the grass with a little of one of these solvents. This time they will find that a green solution is formed and that it will pass through a filter paper. The green solution may now be tested to see if it contains only one green substance or more than one substance. The method used for this is paper chromatography, a method which was introduced in the last section. To get a good separation of colours, careful attention to practical details is necessary.

*Fresh spinach is a good all-the-year-round substitute for grass.

Experiment A1.3

An attempted separation of the green colouring matter in plants by chromatography

Apparatus

Each pupil (or pair) will need:

Experiment sheet 4

Pestle and mortar

Test-tube, 100 × 16 mm

Beaker or evaporating basin as support for filter paper

Filter paper (preferably Whatman No. 1)

Teat pipette

Scissors (unless the grass is already cut up)

Supply of grass (or spinach)

Acetone

Toluene

Procedure

This is described in *Experiment sheet* 4 which is reproduced below.

Experiment sheet 4

You have seen that some kinds of ink contain mixtures of coloured substances. In this experiment the same method will be used to find out if the green colour of plants is due to a single substance or a mixture of substances.

Take a small handful of grass or leaves and cut it into small pieces. Put the pieces into a mortar, add about 3 to 4 cm^3 of acetone, and grind the mixture with a pestle. Add a little more acetone if it seems to be needed. Pour off the liquid into a test-tube.
Has any colouring matter been extracted by the acetone?
What colour is the liquid?

Rest a filter paper on the top of a beaker or evaporating basin. Put one drop of the liquid obtained on to the centre of the paper. Allow it to dry, add another drop and allow this to dry. Repeat until four drops have been added. Now add separate drops of pure acetone, or of toluene, allowing each one to spread before adding the next.
What happens?
Is the green colouring matter in grass a single substance or a mixture?

Allow the filter paper to dry (acetone and toluene are both inflammable).
Draw a diagram of the chromatogram obtained.

If you have time you can try a rather different method of making a chromatogram, using the apparatus shown below. Use four drops of the coloured acetone solution as before, and toluene as the developing liquid.

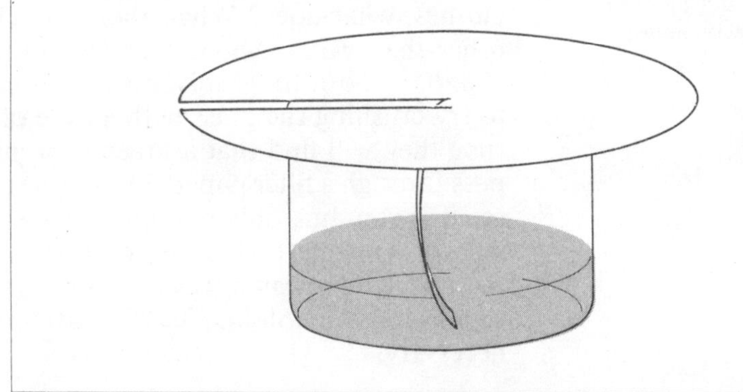

In the chromatogram the outer orange band is xanthophyll and the green band is chlorophyll. There are two types of chlorophyll

(chlorophyll a and b), but these are not separated by this technique. Carotene is also present. An inner ring of carotene can usually be seen if toluene is used as the eluent.

The pupils may take some time to learn how to make good chromatograms, but the time is well spent. Let the pupils keep the chromatograms to stick in their laboratory note books.

When the experiment is over, show film loop 1–2 'Chlorophyll extraction'. This loop shows how chlorophyll, xanthophyll, and carotene are extracted from grass on a large scale. Detailed instructions on how to use the loop are given in the film loop notes.

Finally, if there is time, dip a piece of blackboard chalk (plain white, not the 'dustless' or yellow-covered variety) into the green solution. The appearance of yellow and green rings will show that another type of chromatography is taking place in the chalk. See *Collected experiments* for a fuller description of this experiment.

New words

Chromatography
Chromatogram
Chlorophyll
Xanthophyll

Suggestions for homework

1. Read the Study sheet *Analysis*.
2. Give the pupils pieces of filter paper and tell them to make chromatograms of different inks and dyes at home. They may record these results in their laboratory note books.

Summary

Pupils should now be familiar with the technique of paper chromatography, and have some ability to select the most useful technique for separating an unknown mixture. They should also have a knowledge of the industrial extraction of chlorophyll.

A1.4
Using coloured substances from plants to provide an introduction to acidity–alkalinity

In this section we leave for a time the main theme of the Topic, 'What is a pure substance?', to follow up the extraction of coloured substances from plants and use these coloured substances in an introduction to acidity-alkalinity.

Preliminary work

This section requires that the pupils be prepared one lesson in advance. A short discussion on the coloured matter in plants, flowers, and vegetables leads to the suggestion that each pupil takes home a corked test-tube and brings it back for the next lesson filled with the coloured extract (or juice) of a vegetable, fruit, or flower.

If preparation has been made as suggested, the pupils will arrive for the first lesson in this section with test-tubes containing solutions of coloured substances obtained from fruit, vegetables, or flowers. In

Photographs opposite

A collaborative class exercise. *a.* Pupils are investigating the action of a variety of common substances on Full-range Indicator paper. *b.* Each group is recording the results on a chart as they go along.

case this has not been done, a number of solutions should be prepared before the lesson. A method for preparing such solutions is described below. A wide variety of substances is suitable and the choice will, of course, depend on the time of year and the location of the school. If flowers are scarce, it may be better for the teacher to prepare the coloured extracts in advance. Alternatively flower petals may be collected in summer and dried (an infra-red lamp dries them quickly) and the solution produced when required.

Experiment A1.4a

To prepare solutions of some coloured substances from natural sources

Apparatus

The teacher will need:

Pestle and mortar

Round-bottom flask, 250 cm^3

Liebig condenser to fit flask

Beaker large enough to hold flask

Tripod

Bunsen burner and asbestos square

Filter funnel, stand, and papers

Anti-bumping granules

Plant material

Industrial methylated spirit

The following method gives solutions suitable for the pupils to use in this section. Solutions can be made with, for example, delphinium flowers, blackcurrants, rose petals of various colours, red cabbage leaves, blackberries, and beetroot. Almost any coloured plant material is suitable, with the exception of yellow flowers such as daffodils and dandelions.

Procedure

First crush the plant material with a pestle and mortar and then grind it up thoroughly with a mixture of equal volumes of industrial methylated spirit and water. You will require a quantity of this mixture at least ten times the weight of the plant material being used.

Transfer the material to a round-bottom flask (which should not be more than half filled) and add two or three anti-bumping granules to ensure even boiling. Insert a condenser vertically in the flask, and reflux the contents gently over a boiling water bath until it can be seen that the solid parts of the mixture have become 'white'. This should take about twenty minutes.

Allow to cool and then filter to obtain the coloured extract.

A suggested approach

Objectives for pupils

1. Ability to identify acids and alkalis using an indicator
2. Use of the pH scale to give a numerical value to degrees of acidity–alkalinity
3. Understanding of the meaning of the words: acid, alkali, neutral, indicator

Ask the pupils what they know that has a sharp or 'acid' taste. There are lemon juice, vinegar . . . acid drops. Where else have they come across the word *acid**? They may have seen the advertisements which speak of 'acid stomach' and of a cure in the form of a 'stomach powder'. What does the stomach powder do? It appears from the smiling face on the advertisement that it removes the acidity! The effect of stomach powder on an acid substance can be observed by using the coloured substances from plants. This leads to a search for better ways of showing changes in acidity and to the introduction of an acidity–alkalinity indicator.

Experiment A1.4b

Using some other coloured substances from plants

Apparatus

Each pupil (or pair) will need:

Experiment sheet 5

6 test-tubes, 100 × 16 mm, and rack

(Continued)

Procedure

In this experiment the pupils first test a solution made from an acid drop or from lemon sherbet and then one from stomach powder or Milk of Magnesia. From the acid drop, pupils can move on to other acidic materials, such as lemon juice and vinegar. Full-range

*There is a discussion on the use of the word 'acid' in the *Handbook for teachers*, Chapter 2, pages 29–31.

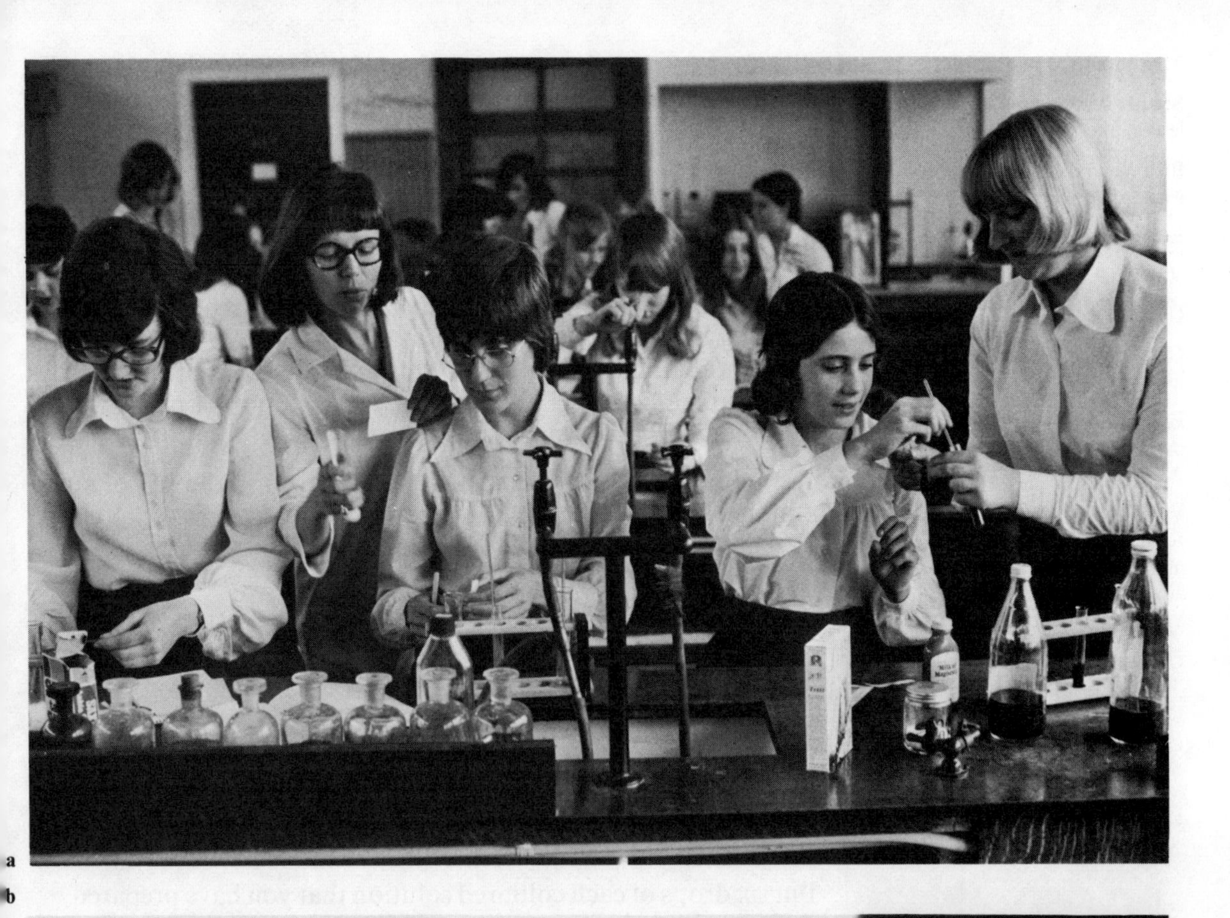

a

b

PH SUBSTANCE
ACID 1
2
3 Vinegar LEMON JUICE
4
5
6 DISTILLED WATER
NEUTRAL 7 Sugar Salt
8 Washing Soda
9
Ammonia
14
N
1-2-3

Teat pipette

Supply of small acid drops, or lemon sherbet, etc.

Proprietary stomach powder or Milk of Magnesia

Lemon juice

Vinegar

Citric acid

Calcium hydroxide (slaked lime)

Extract of plant material prepared as in Experiment A1.4a (or pupils' own extract)

Various acidic and alkaline solutions as available, including 0.1M hydrochloric acid and 0.1M sodium hydrogen carbonate

Full-range Indicator (BDH)

Indicator can then be introduced as a substance that not only indicates whether a substance is acidic but also how acidic it is.

Now they can take a series of solutions and see what happens when Full-range Indicator is added to each solution. How are we to specify the different acidities of these solutions? It would be clumsy continually to refer to the colour of the indicator. We therefore use a scale of numbers – the pH scale. The pupils may be told that there is a quantitative definition which determines this scale, but for the time being it will be used simply as a way of expressing acidity and alkalinity, alkalinity being at this stage simply the property of any solution which has a pH greater than 7. A solution of pH 7, which is the 'acidity' of pure water, is called a neutral solution.

Finally, Full-range Indicator may be used to find the pH of such common materials as toothpaste, lemonade, and tap water. A chart may be made showing the pH of these common substances.

Experiment sheet 5 is reproduced below.

Experiment sheet 5

In this experiment you will study the effect of a number of chemicals on coloured substances extracted from plants. You may have already prepared solutions of some of these coloured substances. If not use the method of *Experiment* 4 with ethanol (alcohol) as a solvent instead of acetone.

Put six drops of each coloured solution that you have prepared in the middle of separate pieces of filter paper and allow them to dry.

While waiting dissolve a small piece of an acid drop or some lemon sherbet in a little water and add four drops of one of your coloured solutions.
What happens?

Now test a solution of an 'anti-acid' substance, such as stomach powder, with the same coloured solution.
What happens this time?

Test each of the substances named in the table overleaf, as follows: add a measure* of the solid substance or ten drops of liquid to about 5 cm^3 distilled water and shake well. Using a teat pipette, place one drop of the solution obtained near the edge of one of your coloured extracts on the filter paper. Make notes of all colour changes in the table. Repeat with other coloured extracts.

*A 'measure' means the amount of solid that would lie on a new penny, or would just fill the grooved end of a Nuffield spatula.

	Extract prepared from		
Colour of extract alone			
Colour with lemon juice			
vinegar			
citric acid solution			
bicarbonate of soda solution			
lime solution			

A more sensitive indicator

Better 'indicators' for acids and alkalis are available. One of these is called Full-range Indicator. This changes colour according to the acidity or alkalinity of the solution. Acidity or alkalinity is given a number (called the 'pH value') according to the colour of the indicator. The lower the number the more acidic the solution, the higher the number the more alkaline. A table will help you to understand.

Colour of Full-range Indicator	Acidity/alkalinity number (pH value)	
Red	1	↑
	2	
Orange	3	increasing acidity
	4	
Orange-yellow	5	
Yellow	6	
Greenish-yellow	7	neutral
Yellowish-green	8	
Green	9	
Bluish-green	10	increasing alkalinity
	11	
Blue	12	
	13	
Violet	14	↓

You can use Full-range Indicator paper to test various liquids and solutions for acidity and alkalinity. Put a small drop of the solution to be tested on a piece of indicator paper and compare the colour with the chart provided with the paper. Record your results in the table below.

Substance tested	Colour with indicator	pH value

At this stage an acid is considered to be a substance which produces an acidic solution and an alkali is a substance which produces an alkaline solution. The definitions of acidic and alkaline are purely operational. Acidic solutions are those with a pH less than 7 and alkaline solutions are those with a pH greater than 7. Pure water and any solution with the same acidity (that is, pH 7) are described as neutral.

New words

Acid (adjective, acidic)
Alkali (adjective, alkaline)
Neutral
Indicator

Suggestion for homework

Supply the pupils with pieces of Full-range Indicator paper to test substances at home, for example, aspirin, detergents, soap, indigestion powders, and Milk of Magnesia. Their findings may then be tabulated.

Summary

Pupils should now know how to use Full-range Indicator to find the pH of solutions. They should also know that solutions having pH numbers less than 7 are called acids; those with pH numbers greater than 7 are called alkalis; and a solution of pH 7 is known as neutral.

A1.5 What can we get from crude oil?

In this section we return to the main theme of the Topic and let the pupils investigate the problem 'Is crude oil a pure substance?' by using a simple form of fractional distillation.

A suggested approach

Objectives for pupils

1. Skill in the technique of fractional distillation
2. Knowledge of the industrial process of crude oil fractionation
3. Understanding of the meaning of the words: fractional distillation and (crude oil) fractionation

The teacher will need

Crude oil sample
(If necessary an artificial crude oil can be made up by mixing petroleum, ether, paraffin, lubricating oil, and a heavier oil)

Film loop projector

Film loop 1–4 'Oil prospecting' (with notes)

Film loop 1–5 'Petroleum fractionation' (with notes)

Show the pupils a sample of crude oil and start a discussion on its importance and uses. The major oil companies produce excellent booklets on this subject.* Emphasize that although oil has been used in the past primarily as a source of energy, it is now extremely important as a source of chemicals. Compounds from crude oil are used for producing a very large number of substances, among them plastics, detergents, anti-freeze (ethylene glycol), and fertilizers. Where and how is crude oil found? This discussion can be usefully supplemented by showing film loop 1–4. Before the lesson, carefully read the notes supplied with the loop.

Now make a careful examination of the crude oil sample. You have said that many substances can be made from crude oil. How can the pupils show whether crude oil consists of a number of substances mixed together or whether it is a single substance? What can be

*The following organizations publish information on the production of materials from crude oil (as well as drilling, distillation, etc.). This information is *only* available to teachers, and they must apply on their school writing paper.

Information Service Department, Institute of Petroleum, 61 New Cavendish St, London WIM 8AR.

Shell Information Service, Shell International Petroleum Company Ltd, Shell Centre, London SE1 7NA.

BP Educational Service, PO Box 21, Redhill, Surrey.

Educational Services, Public Affairs Department, Esso Petroleum Company Ltd, Esso House, Victoria St, London SW1.

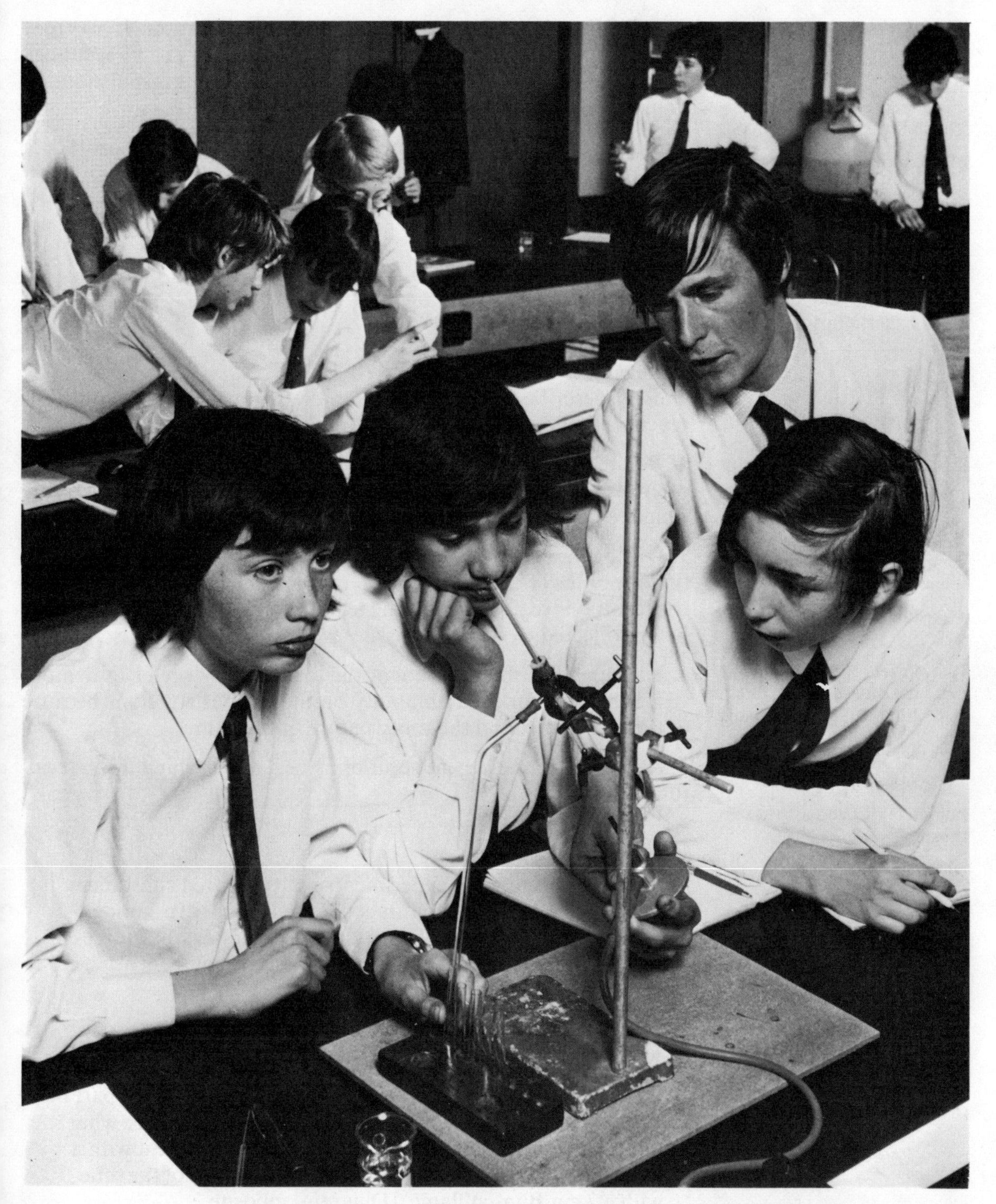

Pupils fractionally distilling crude oil in discussion with their teacher (Experiment A1.5).

Useful films

'Refining'

'Oil'

'North Sea strike'

BBC programme 'Coal age?'

See Appendix 4 for brief descriptions and further details of films and TV programmes

done to find out? The pupils may suggest filtration. If so, they should be allowed to try very little on a filter paper in a filter funnel. Since crude oil is a liquid, they are more likely to suggest distillation or chromatography.

If they suggest chromatography you may let them try a drop on a filter paper or demonstrate it yourself. As this is not successful, distillation will have to be tried. This can be done very simply in the way described below.

Experiment A1.5

Fractional distillation of crude oil

Apparatus

Each pupil (or pair) will need:

Experiment sheet 6

Test-tube, hard-glass, with side-arm ('filter tube'), 125 × 16 mm

Bent delivery tube and rubber connection tubing

4 test-tubes, 75 × 12 mm, in rack

Thermometer, 0–360 °C and cork; the cork should be fitted ready for use to avoid breakage

Teat pipette

4 watch-glasses, hard-glass

Bunsen burner and asbestos square

Stand and 2 clamps

Anti-bumping granules

About 3 cm^3 of crude oil

Procedure

This is described in *Experiment sheet* 6 reproduced below. The teacher should however note the following points.

1. Before the experiment it is wise to try it out yourself, as samples of crude oil vary, and it may be desirable to alter the temperature ranges over which the various fractions are collected.

2. During the experiment pupils' apparatus should be checked to see that the bulbs of their thermometers are in line with the side-arms of the test-tubes so as to record the temperature of the vapour actually being collected. A supply of tissues to mop up spilt oil is useful.

3. At the end of the experiment the teacher may prefer to demonstrate the tests for inflammability on the various fractions because of the risk of fire, and the sooty nature of the flames.

4. A strong, hot detergent solution is best for washing up the test-tubes afterwards.

Experiment sheet 6

You have been doing experiments with coloured substances obtained from plants. In this experiment you will examine crude oil (or petroleum), one of the most important sources of both fuels and chemicals, to see what can be got from it by heating.

Put three or four anti-bumping granules in the bottom of a test-tube. Place about 3 cm^3 of crude oil on them by means of the teat pipette. Be careful to keep the oil off the upper part of the test-tube, as this will make it difficult for you to see what happens later. Clamp the tube at an angle and close it with a cork carrying a thermometer. Warm the bottom of the tube with a 2 cm Bunsen flame. [Diagram opposite.]
What do you notice in the upper part of the tube?

Hold a test-tube to catch any liquid that drips out of the side-arm. When the thermometer reads about 70 °C, collect the liquid in a different test-tube. Again change the test-tube when the temperature has risen to about 120 °C, and again at 170 °C. Collect the last fraction until the temperature is 220 °C. You may need a rather larger flame to drive off the higher-boiling fractions.

When you have finished, you should have collected samples of liquids that boiled over the following ranges:

1. Room temperature to 70 °C
2. 70–120 °C
3. 120–170 °C
4. 170–220 °C

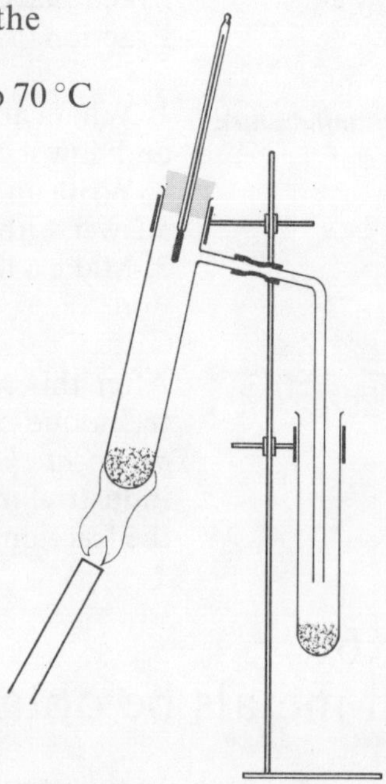

With each of these four test-tubes containing small quantities of materials obtained from the oil, try the following two simple tests:

a. Slowly pour the contents of each test-tube into separate watch-glasses. Compare the ease with which each sample flows. Which flows the most easily?
Which flows least easily?

b. Try to light the liquids in the watch-glasses. You may find it helpful to use a small tuft of cotton wool as a wick. What happens?
Do you think each fraction you have collected is a pure substance?

Pupils could also be told that adding fractions to asbestos paper before placing in a flame gives a good test of flammability.

After the experiment, discuss the results with the class. Tell the pupils that this method of distillation, whereby two or more liquids are separated into fractions (that is, parts of the whole), is called fractional distillation. Have they shown that crude oil is not a single substance?

They will see that they have, but do they think that the fractions they have collected are pure substances? If they were to try to separate them again, using a more sophisticated process, they would find that the fractions contained many different chemicals. Tell them that the process which they have carried out is conducted on a very large scale in industry. Describe briefly how crude oil is fractionally distilled in oil refineries and how, after further processing, the fractions are used for different purposes. Film loop 1–5 'Petroleum fractionation' may be shown at this point. Full details of the film are given in the notes supplied.

New words

Fractional distillation
Fraction (as in crude-oil fraction)

Suggestions for homework *

1. What can you find out about the places where crude oil is found and how it came to be there?
2. Write an account of the uses of fractionation and illustrate your answer with cuttings from magazines and papers.
3. Make a list of things made by processes which involve fractionation.

Summary

After this section the pupils should have some skill in the simple technique of fractional distillation used in Experiment A1.5 (*Experiment sheet* 6), and should know a few simple facts about the industrial methods of separation of crude oil, and the uses to which the fractions may be put.

A1.6
How can metals be obtained from rocks?

In this section a significant step is taken. Up to now the separation of a single substance has always been from a mixture and methods such as distillation, filtration, and paper chromatography have sufficed. Now for the first time they must perform a chemical reaction in order to obtain the pure substance. This raises the subject of mixtures and compounds.

A suggested approach

Objectives for pupils

1. Skill in the technique of heating substances on asbestos paper
2. Recognition that there is a difference between a mixture and a 'chemical compound'
3. Recognition of the use of electricity in the investigation of substances
4. Understanding of the meaning of the words: ore and refining

The teacher will need

One or more samples of a metal ore, including copper pyrites and, if available, a little gold-bearing quartz

Open the subject with a discussion on metals. Ask the pupils how many metals they can name. Do they know where the metals are found? Some metals are found native – that is, as elements. Gold, silver, and copper (rarely) are examples. If gold-bearing quartz is available, show the pupils the ore and point out that the gold is clearly visible. With these metals, the problem of getting the metals is simply one of purification. The only differences between this type of purification and the purifications that they have carried out themselves (salt from rock salt, water from ink, chlorophyll from grass) lie in the scale of the operation and the fact that the metal is usually present in the ore only in very small proportions. (160 tonnes of gold-bearing ore may have to be mined to obtain 1 kg of gold.) The scale of these operations is well illustrated by the film loop 1–7 'Gold mining' which may be shown at this point. Details of the process are given in the notes provided which should be consulted before the lesson.

*See the footnote on page 52 for some sources of the sort of information pupils will need for these homework suggestions.

Some lead carbonate ('cerussite')

Film loop projector

Film loop 1–9 'Copper refining' (with notes)

Film loop 1–7 'Gold mining' (with notes)

Useful film

'Treasure trove'

See Appendix 4 for a brief description and further details

Now explain that most metals are in fact not found native. Show the class a piece of copper pyrites or another similar ore. However closely they look at it under a microscope they will not find particles of metal – and yet it is an ore of copper (in the case of copper pyrites) and copper is obtained from it. Clearly the extraction of this metal is going to be a different problem because the copper is 'contained' in the rock in another way. It is best to let the experiments speak for themselves and then to bring out the idea of a compound during discussion.

The problem is how to extract the metal from an ore such as cerussite. Ask the class for suggestions and let them follow up some of their own ideas if there is time. If help is needed tell them that long ago men found that if the ores of some metals were placed in wood fires the metals were produced. Let them try heating the cerussite on its own and then with carbon. Details are given below.

Experiment A1.6a

How can metals be obtained from rocks? (Part 1)

Apparatus

Each pupil (or pair) will need:

Experiment sheet 7

Bunsen burner and asbestos square

Tongs

Asbestos paper strip about 5 cm × 2 cm

Sample of 'crushed lead ore', e.g. lead carbonate ('cerussite')

Powdered wood charcoal

Procedure

A bead of lead can be obtained from cerussite by heating a mixture of a little crushed ore with charcoal on a piece of asbestos paper. Mix the ore and the charcoal thoroughly and heat with a roaring Bunsen flame.

Experiment sheet 7 is reproduced after Experiment A1.6b.

After this experiment point out that a different technique was necessary to 'release' the lead from cerussite from that used to separate, say, pure salt from rock salt. In the case of the rock salt it would have been quite easy to mix the sandy stuff and the pure salt together again to get something resembling rock salt. But how would you get this lead back into cerussite? We cannot answer the question yet because we do not know enough chemistry, but the point is that for the first time we are in the realm of *chemical* changes. Avoid discussing the changes in terms of oxidation and reduction; see *Handbook for teachers*, Chapter 2, pages 31–32.

Now return to the copper pyrites ore and tell the class that it is a mixture of rocky substances and pure copper pyrites in which copper is combined with sulphur. When copper is extracted industrially the pyrites has to be ground, roasted, and dissolved in sulphuric acid. The copper can be obtained from the resulting solution by using electrical energy. The purpose of the next experiment is to allow the pupils to follow this process on a small scale.

Experiment A1.6b

How can metals be obtained from rocks? (Part 2)

Apparatus

Each pupil (or pair) will need:

Experiment sheet 7

Crushed copper pyrites

(Continued)

Procedure

The teacher should do this experiment first himself, to ensure that the ore really does contain copper; many specimens supplied to schools have been found to consist mostly of iron pyrites.

Details are given in *Experiment sheet* 7 reproduced on the next page.

6 V battery or alternative d.c. supply

2 carbon electrodes with connecting wires and crocodile clips

Beaker, 50 cm^3

Spatula

Bunsen burner and asbestos square

Tongs

Access to filtration apparatus

Pestle and mortar

Asbestos paper strip, about 4 cm × 5 cm

A clean nail

M sulphuric acid

Experiment sheet 7

Metals form another group of important substances that can be obtained from the outer crust of the Earth. You have probably seen one or two pieces of rock from which metals can be obtained. Such a rock is called the 'ore' of the metal. By following the instructions described below you should be able to obtain a little lead and copper from their ores.

1. Mix half a measure of crushed cerussite with a similar volume of charcoal. Place the mixture at one end of a piece of asbestos paper (which should be about 2 cm × 5 cm) and cover it with more charcoal. Heat it in a small but roaring Bunsen flame. When no further change seems to be happening, tip the residue onto a tile or asbestos card.
What do you see?

2. It is more difficult to get the metal copper from its ores, but if you carry out the following directions carefully, you will succeed. Crush a measure of ore in your mortar and heat it strongly on a piece of asbestos paper, double thickness, for a few minutes. Tip the residue into about 5 cm^3 of dilute sulphuric acid in a test-tube and warm gently. Then either filter or centrifuge the mixture. A very pale blue solution should result. To show that this solution contains copper, put a clean iron nail into it.
What do you observe?

Another way of getting copper from the blue solution is to use a battery. Connect the two pieces of carbon rod to an accumulator or other source of electric current and immerse them in the blue solution. After a few minutes, remove the electrodes and look at them. If you see no change, put them back in the solution for another five minutes. Remove them again.
What change, if any, do you observe?

If there is time, replace the electrodes in the solution and leave them for an hour or more. Then examine the solution. Does it look the same as it did at the beginning of the experiment?

Note that it is not intended that the battery should be 'explained' (it is simply a source of energy to be compared with the Bunsen burner).

Some useful points can be made as a result of this experiment. Once again the procedure required to get the copper from the ore was quite different from those processes of separation used in the preceding sections. Point out that the material the pupils finally obtained (copper) looked and behaved quite differently from the ore from which it was obtained. The ink and water could be mixed together again to form a normal ink solution (A1.2), and the fractions of crude oil mixed together again produced something that looked very much like the crude oil they started with (A1.5), but there is no question of producing copper pyrites again by simply mixing substances together. It looks as if neither the copper pyrites nor cerussite can be classed as a mixture. The word compound may be introduced here, but without formal definition, as the pupils will meet many more chemical compounds during Stage I of the course. In doing so they will learn to recognize them through actual experience. Finally, show film loop 1–9 'Copper refining' so that the pupils can see that the reaction they perform in a small beaker is carried out industrially in acres of plant.

New words

Ore
Chemical compound
Refining

Suggestions for homework

1. Read the Study sheet *Where chemicals come from.*
2. Find out the names of as many ores as possible and the names of the metals derived from each of them.
3. Find out the names of as many metals as you can which are obtained by electrical processes.
4. Make a short summary of what you have learnt in this Topic.

Summary

The pupils should now be able to heat substances on asbestos paper, and pass an electric current through a solution; and should have come to regard the Bunsen burner and the battery as two useful weapons in the chemist's armoury for investigating materials. They should have some elementary understanding of the difference between a mixture and a chemical compound, and understand the meaning of the words ore and refining.

Topic A2

The effects of heating substances

Purposes of the Topic

The main purpose of this Topic is to allow the pupils to investigate the effect of heating substances, and to introduce further ideas on the nature of chemical change through observations of 'permanent' and 'temporary' changes. The substances to be heated have been carefully selected and lead the pupils to a number of simple ideas, some of which are followed up in Topics A3 and A4.

Contents

A2.1 What happens when substances are heated?
A2.2 A further look at the effect of heating copper sulphate crystals
A2.3 To find out if there is any change in weight when substances are heated
A2.4 Tracking down the matter lost when potassium permanganate is heated
A2.5 Tracking down the matter gained when copper is heated

Timing

Each of the sections of this Topic will probably need two double periods, except A2.3 and A2.4 which need only one. A total of ten double periods should give plenty of time for discussion. If the time available for this Topic is limited, care is needed to ensure that the real purpose of the Topic is not lost. The number of substances heated in A2.1 may be reduced, or a method evolved whereby each pair of pupils heats only a few substances and results are pooled. Section A2.2 may be omitted as a separate section and combined with A2.5. Section A2.3 could also be shortened. In this way a good class may get through the Topic in four double periods. A slow class might try the reduced course in the standard time (ten double periods).

Introduction to the Topic

To begin the Topic the pupils examine the effect of heat on a variety of different materials. They follow up the heating of copper sulphate crystals and find that water is given off and that the water and the white copper sulphate recombine on mixing.

They then heat a number of substances to see if they gain or lose in weight. This leads to a closer examination of what happens when potassium permanganate loses weight on being heated and copper gains weight on being heated, and they find that a gas is given off in the first change and that air is needed for the second change.

By the end of the Topic the pupils will have met a variety of new substances and they will have learnt how to cope with some new techniques, notably the collection of gases over water or with syringes. They will have considered very briefly the states of matter and they will be on the way to understanding the difference between an element and a compound. Note that the concept of an element is implicit but not explicit in this Topic. Similarly they will be able to recognize with growing confidence when a chemical change is taking place. In the last two sections they will again have developed their powers of reasoning by applying them to problems in chemistry.

Background knowledge

Pupils will come to this Topic with an understanding of what is meant by a single substance, and it is here assumed that investigations are best carried out on such single substances. Pupils should also know how to use a Bunsen burner, and the usefulness of boiling points when attempting to identify a liquid.

Subsequent development

The ability to put forward theories and test them by experiments will be used again and again in later Topics. In particular, the results of the experiments in which copper, and potassium permanganate, are heated are followed up in Topic A3. In Topic A3 these results, and ones from further experiments, are used in studying the composition of the air. The subject is continued in Topic A4 'The problem of burning'.

Further references

for the teacher

Further experiments suitable for Topic A2 are to be found in *Collected experiments*, Chapter 1 'The effects of heating substances'.

The film 'Exploring chemistry' deals with the investigational approach to chemistry teaching, and part of it includes a lesson on section A2.5. See Appendix 4 for a brief description and further details of the film.

Reading material

for the pupil

Study sheet:
Heating things shows that when pupils use heat from a Bunsen burner they are doing what, on a much larger scale, is done in industry every day to make such familiar materials as glass, china, and steel. Cooking is also given as an example of how heat can change a substance into something more useful.

A2.1
What happens when substances are heated?

The first section in the Topic is an exercise in observation. The pupils learn how to heat things and to observe carefully what changes take place. The introduction of the idea of temporary and permanent changes leads them to think again about the nature of chemical changes. (See A1.6 where the idea was first introduced.) No distinction is made between 'physical' and 'chemical' changes as such and the pupils are introduced from the start to reversible chemical changes. The section leads on to further qualitative and quantitative investigations in A2.2 and A2.3.

A suggested approach

Objectives for pupils

1. Skill in the technique of heating substances in test-tubes
2. Ability to classify
(Continued)

Tell the pupils that when a chemist investigates unknown substances, he tries to bring about changes in them. In doing this he has a number of lines of attack – he can mix 'this' with 'that' and see what happens; he can heat things and see what happens; he can try to pass an electric current through things; and so on. In this Topic, we shall use one of these methods of attack, namely, the action of heat.

Ask the class for examples of things that change when you heat

experimental results, and to distinguish between temporary and permanent changes

them. There is ice, for instance, which changes to water when it is heated, and water which turns to steam. Now tell them that the object of their next experiment is to find out what happens when a number of substances are heated. When they heat these substances, they should observe carefully what happens to them and record the changes. It will be interesting for them to see if there are any patterns in the behaviour of these substances. They should look out particularly for substances which change only temporarily when they are heated, and for substances which appear to undergo a permanent change. Of those that change permanently there are two types: those which can be re-formed by mixing the products together again, and those which cannot. This division should appear from the pupils' experiments (see details below) and may therefore be left until the results of the experiments are discussed.

Experiment A2.1

What happens to some chemicals when they are heated?

Apparatus

Each pupil (or pair) will need:

Experiment sheet 8

Bunsen burner and asbestos square

6 hard-glass test-tubes, 75 × 12 mm

Tongs
Test-tube holder

Iodine crystals

Zinc oxide

Copper(II) sulphate crystals

Cobalt chloride crystals

Magnesium ribbon, about 2 cm

Copper foil, a 2 cm square

Potassium permanganate

Procedure

Let the class heat a variety of substances and observe a number of changes, one or two of which can be followed up in more detail shortly. The substances chosen are grouped into:

1. Those that return to their previous state on cooling. Possible substances are iodine crystals and zinc oxide.
2. Those that split up but can be re-formed by putting together the products that resulted from heating them – copper sulphate crystals, cobalt chloride crystals.
3. Those that seem to have changed irreversibly – magnesium ribbon, copper foil, potassium permanganate.

The pupils should not be told of this grouping in the first instance.

Warning. Iodine crystals attack the skin and the vapour is poisonous. Use one small crystal only for the experiment. It is a good plan for the teacher to have test-tubes containing an appropriate amount ready before the experiment, as this avoids iodine stains being found later on books, clothing, etc.

Each substance can be put out on a piece of paper and numbered, and the pupils told to take a spatula full of each in turn. Most of the substances are easily examined by heating in a 75 × 12 mm test-tube, held in a test-tube holder, and heated with a 3 cm Bunsen flame. The magnesium ribbon and copper foil should be held in tongs and heated directly by the flame. The Bunsen burners should be on asbestos squares.

Experiment sheet 8

So far you have been mainly occupied in separating mixtures of substances. You are now starting a new section in which you will study some chemicals more closely. First you will see what can be found out by heating certain chemicals. These have been chosen because they lead to interesting results which are not too difficult to explain. When heated, some solid substances melt to form a liquid (which may boil) and return to their original state on cooling. Other substances change to new substances.

There is quite a lot to record in this experiment so make a table for this on a separate sheet of paper which can be fixed into your folder. Use these headings:

Name of substance	Appearance		
	before heating	during heating	after cooling

For powdered solids use one measure in a test-tube, with a Bunsen burner flame which is non-luminous and about 5 cm high. Hold the test-tube in a holder (a neatly folded piece of paper can be used) and heat the lower end. Metals (such as magnesium ribbon and copper foil) should be held in tongs and heated directly in a flame.

Be very careful with magnesium. Hold it steadily and do not look at it directly when it burns as you may hurt your eyes.

Which of the substances used change permanently when heated?
Which of the substances seem to be the same after cooling as they were before heating?
What else can you say about this experiment?

After the experiment, ask the pupils what happened in each case. Did they find any pattern in the behaviour of the substances? Did they find any substances which did not change at all? Which substances changed temporarily and which permanently? Some of them may have noticed that there were two ways in which the substances could change permanently. An example of re-forming the original substance is mixing white copper sulphate with the colourless liquid that condensed on the test-tube wall. It goes back to its original colour. It is worth drawing attention to this experiment as it will be followed up in the next section. They may assume that the liquid is water. If so, ask them how they can demonstrate that it is water. The answer to this question is to determine the boiling point, as explained in section A1.2.

Suggestions for homework

1. Read the Study sheet *Heating things*.
2. What other common materials change when they are heated?
3. Think of as many changes in substances being heated as you can, and divide the substances into three groups: those that change back to their original state on cooling, those that split up but can be re-formed by putting together the products that resulted from heating them, and those that change but cannot be changed back.

Summary

Pupils should now have some skill in heating substances, observing the changes that take place, and classifying their observations. It is not intended that they should necessarily remember *all* the results; but they should know the three important ones that are followed up in later sections.

A2.2
A further look at the effect of heating copper sulphate crystals

One of the simple observations of the last section is now put under closer scrutiny. From 'What did you see?' we move on to 'What happened?' In the simple investigation which follows the pupils are asked to give evidence for their assumptions. 'It looks like water but how can you demonstrate that it is water?' For the first time the attention of the pupils is drawn to the energy which accompanies chemical changes.

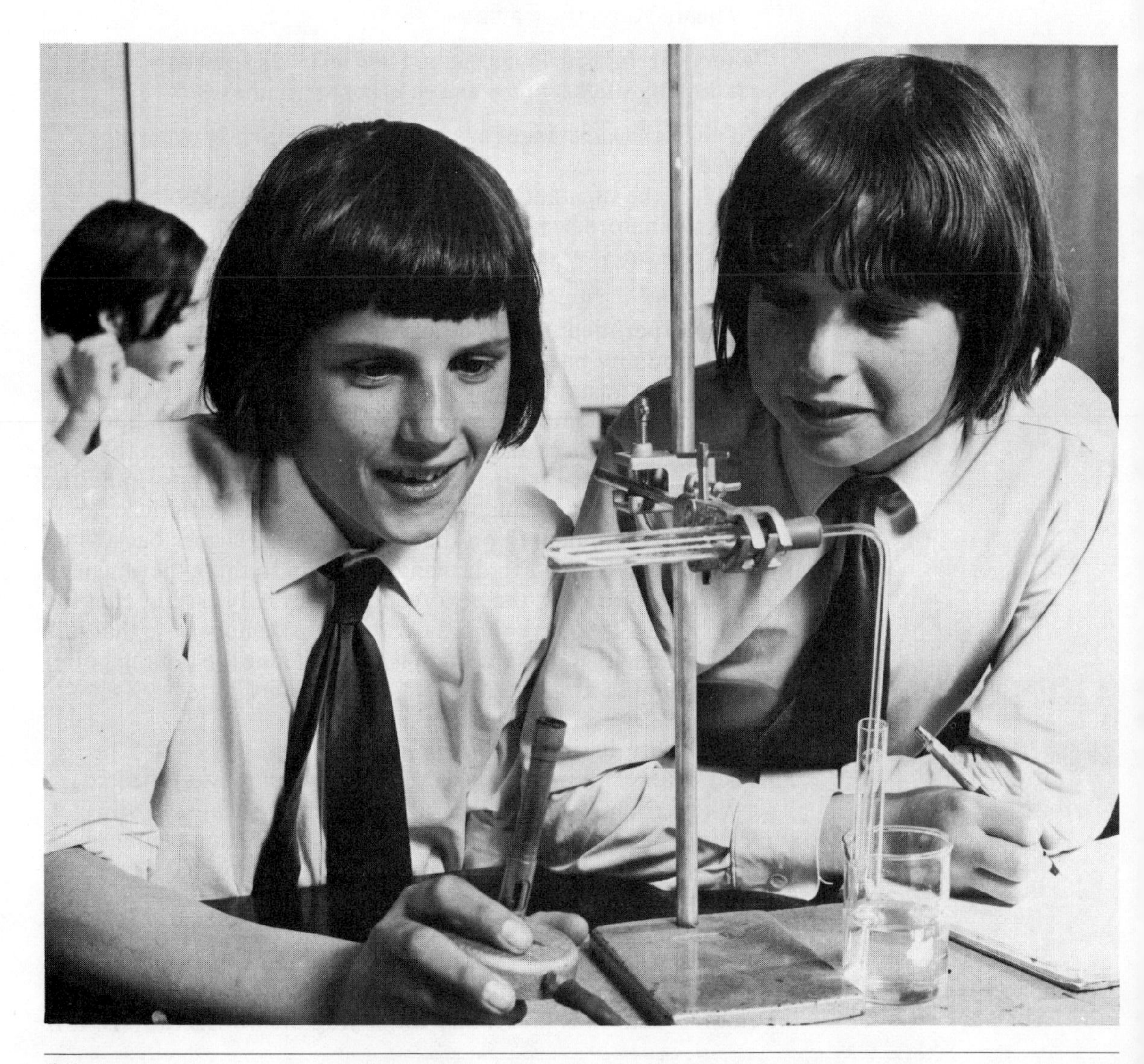

A suggested approach

Objectives for pupils

1. Awareness that assumptions must be tested by experiment
2. Awareness that energy changes accompany chemical changes
3. Awareness of the possibility that some chemical changes can be reversed
4. Understanding of the meaning of the words anhydrous and decompose

Remind the pupils how they heated copper sulphate crystals (Experiment A2.1). They should have noticed that, when the blue copper sulphate was heated, a colourless liquid condensed on the walls of the test-tube. The first question to be asked is 'What is the colourless liquid?' Most pupils will assume that it is water but this assumption must be tested, for example by determining the boiling point as explained in section A1.2.

If the boiling point of the liquid which we suspect is water is to be measured, an apparatus must be devised to collect the liquid. Ask for suggestions from the class and build up an apparatus similar to the one suggested below. The pupils may now try the experiment themselves. Details are given below.

Experiment A2.2

Apparatus

Each pupil (or pair) will need:

Experiment sheet 9

2 hard-glass test-tubes, 100 × 16 mm

Cork or bung to fit test-tube, carrying a delivery tube

Bunsen burner and asbestos square

Thermometer, −10 to +110 °C

Stand and 2 clamps

Beaker, 100 cm^3

Copper(II) sulphate, small crystals

Investigating the effect of heat on copper sulphate crystals in more detail

Procedure

When the instructions given in *Experiment sheet* 9 have been carried out, let the pupils see the effect of adding several liquids to anhydrous copper sulphate (examples might be hexane, tetrachloromethane, acetone, and water) and thus introduce the use of anhydrous copper sulphate as another test for water.

Experiment sheet 9

Having observed what happens when a number of substances are heated, you can now follow up one of these changes more carefully.

What happened when you heated a few crystals of copper sulphate in a previous lesson?

In this experiment you are going to try to find out more about this change.

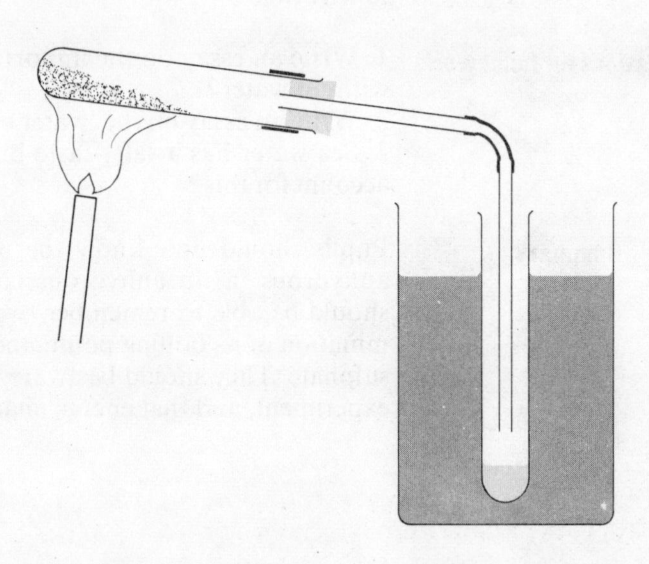

Experiment sheet 9 (continued)

> Half fill a test-tube with dry blue crystals and arrange the apparatus as shown in the diagram [previous page]. Gently heat the crystals with a medium Bunsen flame.
> What do you see happening to the crystals?
>
> Carry on with the heating until you think the reaction is complete. What do you think you have in the righthand test-tube?
> How could you test your suggestions?
>
> Remove the righthand test-tube and clamp it almost vertically. Heat the contents with a small Bunsen flame. Hold a thermometer in the test-tube so that the bulb is near the top and keep it there for a minute or so.
> What temperature does it register? . . . °C.
> What information does this give you about the contents of the tube?
> Do you think the crystals have gone for good? How might you try to re-make them?
>
> Try to do so, using only a small amount of the residue left in the test-tube.
> What *two* things do you notice happening to the residue?

After the experiment the results should be discussed with the class. Are they satisfied that the liquid which came off was water? Draw attention to the energy changes which took place. They had to put heat energy in to split the copper sulphate into white copper sulphate and water. When the two substances were mixed together again they got some of the heat back ('It got hot'). Explain that the white copper sulphate is called anhydrous ('without water') copper sulphate. Crystalline substances containing water are said to be hydrated.

New words

Decompose
Anhydrous

Suggestions for homework

1. Write an essay on the importance of water: 'What would we do without water?'
2. Write an essay on the 'water cycle'.
3. Sea water has a salty taste but rain water has not; how do you account for this?

Summary

Pupils should now know the meaning of the words decompose, anhydrous (as in anhydrous copper sulphate), and hydrate, and should be able to remember two tests for water, namely (1) determination of its boiling point, and (2) its effect on anhydrous copper sulphate. They should be aware that assumptions must be tested by experiment, and that energy changes accompany chemical changes.

A2.3
To find out if there is any change in weight when substances are heated

The emphasis now changes from qualitative work to quantitative work. Significant advances were made in chemistry only when chemists, notably Black and Lavoisier, began to use the balance. In this section therefore we ask the pupils for the first time 'How much?' The measurements of the changes in weight* when substances are heated are then used to find out what is happening. This is the first stage in the search for an explanation of the phenomenon of burning.

A suggested approach

Objectives for pupils

1. Awareness of the value of quantitative experiments
2. Familiarity with the technique of weighing

The effects of heating substances have been examined qualitatively. Point out to the pupils that there are limits to what we can learn about matter by merely observing its behaviour. In the early days of scientific investigation some of the most important discoveries were made when scientists began to make measurements. A brief reference to Boyle, Black, and Lavoisier will emphasize this point. Lavoisier weighed substances before and after heating them. Tell the pupils that they should do the same. The balance may now be introduced as an essential tool in chemical research. The theory of weighing will be left to a physics lesson. To the pupil the balance is a device for finding the weight* of a substance quickly and accurately. For this purpose we recommend a direct reading balance.

Experiment A2.3

Apparatus

Each pupil (or pair) will need:

Experiment sheet 10

Small crucible and lid

Tongs

Pipe-clay triangle

Bunsen burner, tripod, and asbestos square

Use of laboratory balance capable of detecting changes of 0.005 g

Hard-glass test-tube, 100 × 16 mm

Test-tube holder

Magnesium ribbon

Copper, made by reducing wire-form copper(II) oxide

Nichrome wire

Potassium permanganate

To find out if there is any change in weight when certain substances are heated

Procedure

The substances chosen for this experiment are magnesium, copper, potassium permanganate, and nichrome wire. Pupils' attention should be drawn to the instruction in *Experiment sheet* 10 to heat the potassium permanganate very gently. Vigorous heating will cause loss of solid from the mouth of the test-tube.

Experiment sheet 10

In this last experiment when blue copper sulphate was heated, it gave off something. If you had heated it in an open dish, you might not have noticed this. But had you *weighed* it before and after heating, what do you think you would have found? The balance is a very important piece of apparatus in the study of science. In this experiment you are going to use it to find whether there is any change in weight when certain other substances are heated.

1. *Magnesium*. Make a tight coil of the magnesium ribbon, place it in the crucible, put the lid on, and weigh the whole. It weighs . . . g

*The term *weight* is used in Stage I of this course, but some teachers may prefer to use the term *mass* (as in Nuffield Combined Science), especially if the pupils are following a Physics course in which *mass* is used.

Experiment sheet 10 *(continued)*

Arrange to heat the crucible and contents as shown in the diagram.

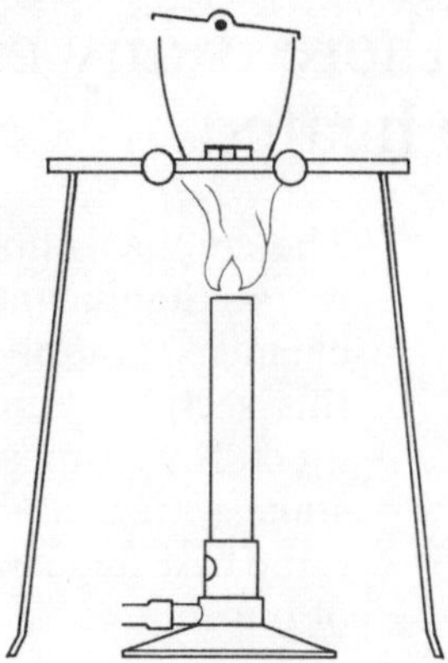

Start with a medium Bunsen flame. Grip the lid with crucible tongs, and move it to one side to see what is happening to the magnesium. If it is not burning, heat it more strongly until it catches fire, but keep the lid on most of the time so that as little smoke as possible is lost.

When the reaction is over, allow the crucible to cool, and when it is cool enough to touch, weigh it again. What is the weight? Has there been a change in weight?

2. *Copper*. Cut up the copper foil you have been given or use 'wire-form' copper. Put it in a crucible and weigh it (with no lid this time).
The weight is . . . g

Arrange to heat the crucible and contents as before. Heat gently for about half a minute, and then strongly for at least five minutes. Wait until the crucible is cool enough to touch and weigh it again.
What is the weight now?
Has there been a change in weight?

3. *Potassium permanganate*. Put two measures of potassium permanganate into a small hard-glass test-tube. Weigh tube and contents. The weight is . . . g

Hold the test-tube nearly horizontally (mouth of tube upwards) in a test-tube holder and warm it *very gently*, in a small Bunsen flame, for about two minutes. Allow to cool and weigh. The weight now is . . . g
Has there been a change in weight?

4. *Nichrome wire*. Use the same methods as for copper, (2) above.
Weight of crucible + wire before heating . . . g
Weight of crucible + wire after heating . . . g
Has there been a change in weight?

Make a summary below, in the form of a table, to show the conclusion from each experiment.

After the pupils have completed the experiment collect the results together on the blackboard and discuss them. Magnesium and copper will have been found to gain in weight, potassium permanganate to have lost weight, and nichrome wire not to have changed. Why is this? If something is lost, where has it gone? If something is gained, where has it come from? The pupils will make suggestions, some of which may be followed up experimentally. Lead them to the idea that they can find out what happens to the substance lost when potassium permanganate was heated. How will they do it? They can think this problem over before the next lesson.

Suggestions for homework

1. Read your own notes on the experiments covered in this Topic. What are the common features? List them. Are *all* changes of the same kind?
2. Make suggestions on ways in which you could find out, by experiment, what happens to the substance lost when potassium permanganate is heated.

Summary

Pupils have heated several substances, weighing them before and after heating, and should now know how to weigh, and appreciate the value of quantitative experiments. Two of the results will be followed up in later sections.

A2.4
Tracking down the matter lost when potassium permanganate is heated

The pupils found during the course of the last section that potassium permanganate loses weight on heating. Now we ask the question 'What happens to the stuff that is lost?'

During the experiment designed to answer this question the pupils learn how to collect a gas and learn how to test for the 'active gas' that comes off.

A suggested approach

Objectives for pupils

1. Development of the idea of a compound
2. Awareness of the existence of gases other than the air
3. Familiarity with the techniques used for collecting gases

Remind the pupils of the lesson in which they found that potassium permanganate loses weight on heating. Ask them what they think has happened to the stuff that has gone. Did they see anything go? If not, does this mean that the stuff that went was a colourless gas? This is one explanation. Ask them how they could test it. They must find a way in which a gas can be collected. They may find it difficult to think of ways of doing this in practice, in which case it will be necessary to show them how to collect a gas over water. A neat alternative is to use a glass (or plastic) syringe. If one is available show them this too. If this is attached to the end of a test-tube (see diagram *a*) it is easy to see whether a gas is given off or not and, if there is, how much gas is produced. Discuss how the apparatus should be set up and let the pupils start the experiment. The gas is not named oxygen until the next section.

Experiment A2.4

Tracking down the matter lost when potassium permanganate is heated

Apparatus

Each pupil (or pair) will need:

Experiment sheet 11

Bunsen burner and asbestos square

Stand and clamp

Hard-glass test-tube, 100 × 16 mm

Wood splints

Potassium permanganate

Glass wool

Further apparatus needed will depend on which method of collection is used:

a. Collection by syringe

Syringe, glass, 100 cm^3

Stand and syringe holder

Small curved delivery tube fitted with bung and connection tubing

b. Collection over water

Delivery tube fitted with bung and connection tubing

Small trough

2 test-tubes, 150 × 25 mm, and corks

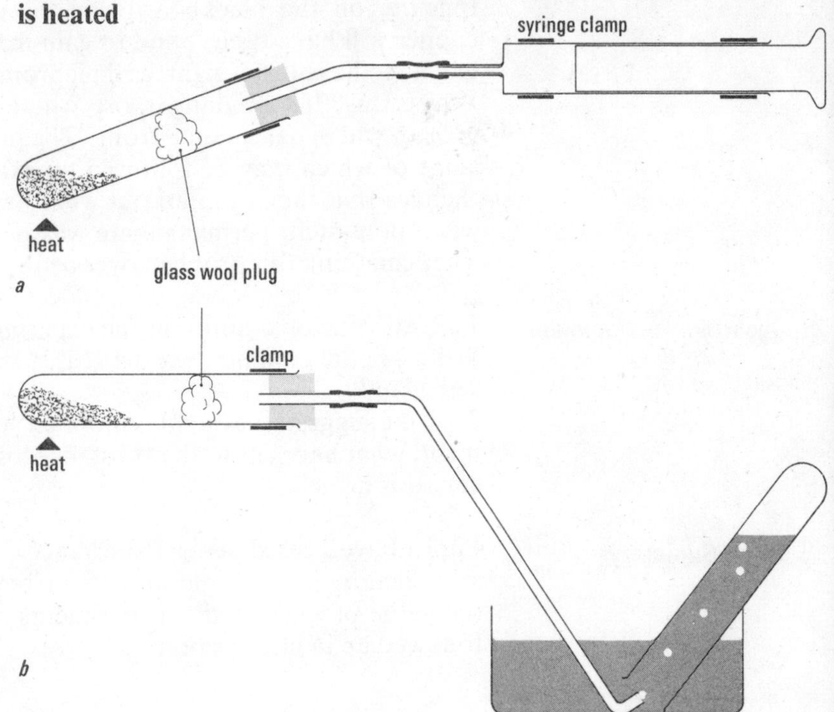

Procedure

Much will depend here upon the apparatus available, and the suggestions made by the pupils. Two possible experimental procedures are now described; *Experiment sheet* 11 is written to suit either, and is reproduced after the procedures have been described.

a. Collection by syringe. About a 5 mm depth of potassium permanganate is placed in the 150 × 25 mm hard-glass test-tube which is clamped at an angle of about 30° to the horizontal and fitted with a loose glass wool plug near its mouth and a short curved delivery tube. The syringe is joined to this by a short piece of rubber tubing and is clamped lightly in a horizontal position. The piston is pushed right in at the beginning of the experiment and is gradually pushed back as gas is evolved from the heated potassium permanganate. The test-tube should be warmed with a very small Bunsen flame which must be moved about to prevent more than a gentle heat reaching any part of the test-tube. The volume of gas evolved can easily be measured.

b. Collection over water. Place a 5 mm depth of potassium permanganate in a 150 × 25 mm hard-glass test-tube, insert a loose plug of glass wool and the delivery tube as described in (*a*), and clamp the test-tube with the end of the delivery tube about 1 cm below the surface of some water contained in a small trough. Gently warm the potassium permanganate by means of the Bunsen burner as described before.

When about half a test-tube of gas (which will be largely displaced air) has been collected, collect a further two test-tubes full of the gas by displacement of water.

When sufficient gas has been collected the end of the delivery tube must be taken out of the water before heating is stopped, otherwise water will be drawn into the test-tube and cause it to crack.

When the gas has been collected ask the pupils to find out if it is like air in supporting the combustion of a burning splint. This may be done by gently expelling some of the gas from the syringe or by using one of the test-tubes full of gas. Then suggest using a glowing splint in the same way so that they may see the splint rekindle in the remaining sample of the gas.

Experiment sheet 11
In the last experiment you found that potassium permanganate lost weight on being heated. You are now going to try to find out whether any invisible substance escaped and whether it can be collected.
Draw a diagram of the apparatus you are going to use.

Use three measures of potassium permanganate and heat it very gently using a small Bunsen flame.
What do you see happening?
Does the gas given off look like air?
Collect a test-tube full of the gas and place a burning splint in it. What happens?

Now use a glowing splint in the same way in a second test-tube full of the gas. What happens?
What do you think the gas is?

After the experiment discuss the results with the class. It should be quite clear that a gas was given off and that this is why the potassium permanganate lost weight. What gas was it? It was certainly like air in that things burnt in it, but it was more 'active' than air in that the glowing splint burst into flame when it was placed in the gas. It may be necessary to repeat this test in front of them.

Suggestion for homework

Refer to your results of Experiment A2.1. You will see that each material behaved differently. What other experiments could you try, to find out what happens when (*a*) magnesium, or (*b*) zinc oxide, is heated?

Summary

Pupils should now know how to collect gases evolved in the course of a chemical reaction. They should be aware of the existence of gases other than the air, but so far should not have been told the name of the gas obtained by heating potassium permanganate.

A2.5
Tracking down the matter gained when copper is heated

In the last section of this Topic the idea is to lead the pupils to find out that air is needed for copper to gain weight when heated and thus that the copper has 'taken' something from the air. Once again the pupils are learning to make deductions strictly from the evidence that they see before them and to plan experiments to test those deductions.

A suggested approach

Objectives for pupils

1. Development of the idea of a compound
2. Ability to speculate, and design experiments to test theories

If the homework in the last section was set, discuss the answers. Then ask the class what happened when they heated copper in section A2.3. Did it gain or lose weight? It gained weight and they will have noticed that, at the same time, it turned black. If they tried scraping off the black stuff they would find that the copper underneath was unchanged. Now ask the class how they think the copper came to gain in weight. Where did the extra stuff come from? They may suggest that it came from the air, from the flame, or from some other source. How can they show which theory is true? The flame theory may be tested by heating the copper in a crucible. It still turns black. How, then, can they see whether it is contact with air that is necessary? Heating the copper in the absence of air will test this point. There are several ways of doing this described below. The film for teachers, 'Exploring chemistry', illustrates the points raised.

Experiment A2.5

Apparatus

The teacher will need:

Bunsen burner and asbestos square

Soft-glass test-tube, 100 × 16 mm

2 hard-glass test-tubes, 100 × 16 mm

Bung and delivery tube to fit hard-glass test-tube

Hammer

Vacuum pump

Pressure tubing to connect delivery tube to vacuum pump

Stand and clamp, or test-tube holder

Tongs

Safety screen

Copper foil

Heating copper away from the air

The first two parts of the experiment are to be done by the teacher to avoid accident.

Procedure

There are several ways of heating copper away from air; in each of following a piece of copper foil about 2 cm square is suitable.

1. Place the foil in a 100 × 16 mm soft-glass test-tube with a few drops of water and place a cork *loosely* in the mouth of the test-tube. Hold the test-tube in a clamp, or by means of a test-tube holder, and heat it gently until all the water has become steam and the air has been displaced. Next, push the cork in firmly, while continuing to heat the tube to redness. The glass may collapse onto the copper but, whether protected by steam or glass, the copper is not exposed to the air and does not turn black.

2. Place the foil in a 100 × 16 mm hard-glass test-tube fitted with a bung and delivery tube attached to a vacuum pump. Place a safety screen between apparatus and pupils, and wear safety glasses. Pump out the air from the tube and heat it to redness. On cooling it will be seen that the copper has not acquired a black coating.

3. Make an 'envelope' out of a piece of copper foil by folding it over itself and flattening the edges with a hammer. Heat the envelope by holding it in a Bunsen burner flame with a pair of tongs. Allow it to cool and then open it out. The inside remains unblackened. This experiment is suitable for the pupils.

Suggestion for homework

Make a summary of the work you have done in this Topic and review the experiments performed. Say why they were done and what the results mean. Suggest further experiments which will help you to find out more about the air.

Summary

Pupils have had a further opportunity of putting forward suggestions and testing them by experiment. They should now realize that the black coating formed when copper is heated is in some way connected with the air. This point is taken up in the next Topic.

Topic A3

Finding out more about the air

Purpose of the Topic

The object of this Topic is to lead the pupils, through their own experiments, to the idea that the gas given off when potassium permanganate is heated is the same as the 'active' part of the air (oxygen) which reacts with copper when it is heated in air.

Contents

A3.1 Is air used up when copper is heated in it?
A3.2 The history of the discovery of oxygen

Timing

Three double periods should be enough to cover this Topic. A3.1 may need a little more than a double period.

If A3.2 is covered in one double period, this leaves a double period over for further discussion and experiment if necessary.

Introduction to the Topic

The Topic begins with an experiment which shows that air is used up when copper is heated in it. The remaining gas is tested and is found to be inert. The Topic ends with a description of Priestley's and Lavoisier's work on oxygen and the nature of combustion, and the gas produced when potassium permanganate is heated is then identified by some of its properties as the 'active' part of the air, oxygen.

By the end of this Topic the pupils should appreciate the experimental evidence, outlined above, for believing that air is a mixture of oxygen and nitrogen. They will also have learnt more about the handling of gases.

Background knowledge

Pupils will come to this Topic knowing (1) that potassium permanganate when heated loses weight and evolves a gas which rekindles a glowing splint; (2) that copper when heated becomes covered with a black coating and gets heavier, except in the absence of air.

Subsequent development

Knowledge gained in this Topic is used in Topic 4, where the part that air plays in burning is investigated.

Further references
for the teacher

Additional experiments on this Topic are given in *Collected experiments*, Chapter 3 'Air and combustion'.

The background to the subject of the discovery of oxygen and the nature of combustion is well documented. For authoritative accounts see Partington, J. R. (1962) *A history of chemistry*. Macmillan. McKie, D. (1952) *Antoine Lavoisier*. Constable.

Many of the early experiments important in the discovery of oxygen and of the role that it plays in combustion are recreated in the film 'History of the discovery of oxygen'. See Appendix 4 for a brief description and further details of this film.

Supplementary materials

Film loop
1–6 'Liquid air fractionation' and chart

Films
'History of the discovery of oxygen'
'O for oxygen'
BBC programme 'Oxygen'
See Appendix 4 for brief descriptions and further details.

Reading material *for the pupil*

Study sheet:
Burning and Lavoisier starts with early ideas about burning, including those of the phlogistonists. Lavoisier's great contribution to our understanding of the processes of combustion and breathing is simply explained and set in the context of his life and times.

A3.1
Is air used up when copper is heated in it?

The pupils now know that for copper to be blackened, heating in air is necessary. They know too that copper gains weight when it is heated. But is air used up when copper is heated in it? The experiments which the pupils devise teach them more about testing ideas and how to handle gases.

A suggested approach

Objectives for pupils

1. Knowledge that the air consists of two parts, one of which is responsible for the formation of the black compound with copper
2. Ability to speculate and design experiments to test theories
3. Familiarity with the quantitative handling of gases
4. Understanding of the meaning of the word inert

In the preliminary discussion go over suggestions made in the pupils' homework for section A2.5 and remind them that they found air is necessary for the blackening of copper.

They used syringes (or another technique) to show that potassium permanganate gives off a gas when it is heated. Can they think of a way in which syringes could be used to find out if air is used up when copper is heated in it? After discussion let them try the experiment described below or, if syringes are in short supply, demonstrate it yourself.

Experiment A3.1

Apparatus

a.

2 stands and syringe clamps

2 syringes, glass, 100 cm^3

15 cm of 7 mm o.d. transparent silica tubing

Bunsen burner and asbestos square

Thick-walled rubber tubing

3-way stopcock in capillary tubing (optional)

(Continued)

How much air is used up when copper is heated?
This experiment may be done by the teacher, or by the pupils if enough syringes are available.

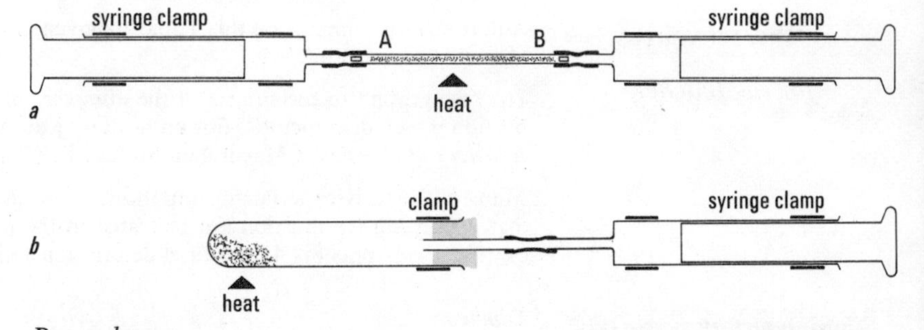

Procedure
In this experiment some copper (preferably made by reducing wire-form copper(II) oxide in a stream of hydrogen from a cylinder of the

2 pieces of hard-glass rod, 5 mm diameter

Copper made by freshly reducing wire-form copper(II) oxide

or *b*.

2 stands and clamp

Syringe, 100 cm^3, and syringe clamp

Hard-glass test-tube, 150 × 25 mm, fitted with bung and straight delivery tube

Rubber connecting tubing

Bunsen burner and asbestos square

Spatula

Vaseline

Copper made by freshly reducing wire-form copper(II) oxide

gas) is placed in a horizontal silica tube, between two syringes. There must be no gaps between the syringes and the tube. Fill in the ends of the combustion tube with pieces of hard-glass rod which are slightly smaller in diameter than the inside diameter of the tube (see diagram *a*).

The experiment may be performed with only one syringe if the apparatus shown in diagram *b* is used. In this case the syringe should be lightly greased with vaseline and contain about 30 cm^3 of air initially. Two or three spatula measures of reduced wire-form copper(II) oxide are needed in the test-tube.

The copper is put in the tube with pieces of the glass rod placed at either end. One syringe is then filled with air. *At this point the apparatus must be tested for airtightness.*

The copper at A (diagram *a*) is now heated fairly vigorously, and when it is hot the syringe plungers are moved slowly backwards and forwards while the heating is continued. After about two minutes have elapsed stop the heating and cool the combustion tube with a damp cloth. Note the volume reading on the syringe again. Repeat the heating of the copper at A for a further minute, cool as before, and continue this procedure until the volume reading is constant. Now use the Bunsen burner to heat the copper at B (see diagram *a*).

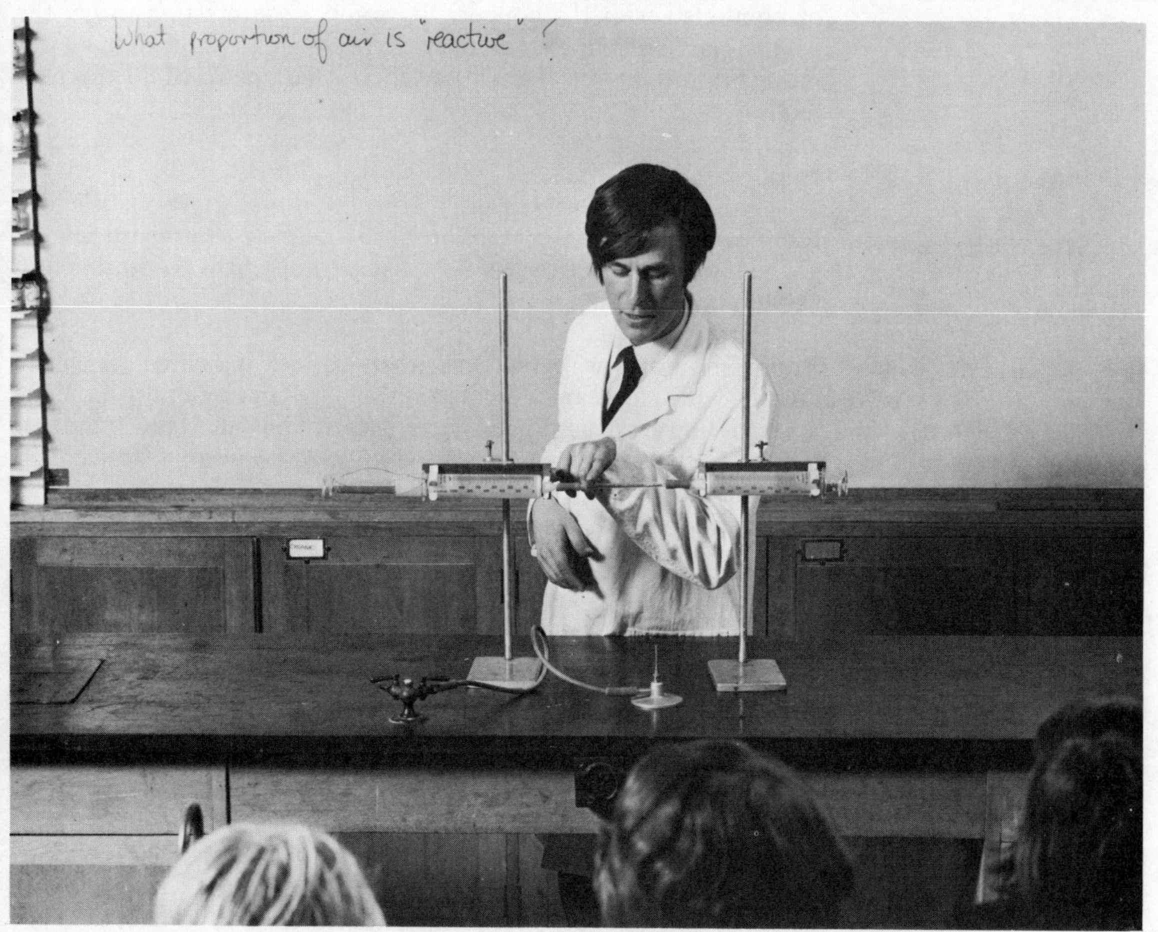

Ideally the copper at B will remain bright and untarnished, and the volume reading constant, indicating that the active part of the air has been removed completely. The volume occupied by the active part of the air is the difference between the first and the final volume readings. From these two volumes the percentage of the 'active part' of the air can be calculated. Finally test the gas remaining in the syringe by expelling it gently over a glowing splint. The splint will go out. Compare this with the effect of passing the gas evolved from potassium permanganate over a glowing splint (see A2.4).

After the experiment discuss the results with the class. If they have done the syringe experiment they will have seen that for every 100 cm^3 of air about 20 cm^3 was 'used up'. The remaining gas in the syringe was 'inactive' or inert. In the alternative experiment (*b*), using one syringe only, after heating the copper for about two minutes a definite decrease in the syringe reading will be apparent when the test-tube is allowed to cool. However, in this case the 'inactive' gas will be in the test-tube, rather than the syringe; less accurate quantitative results are to be expected in this method. Thus the gas absorbed was the 'active' part of the air. Ask then where the gas has gone. If they weighed the copper before and after the experiment, they will know that it gained in weight (they knew this also from experiment A2.3), so that it looks as if the gas has 'gone into' or combined with the copper. 'It would be interesting to get the "active part" of the air on its own.' 'If we did, would things burn more brightly in it?' This question will be answered in the next section.

New word

Inert

Suggestion for homework

How could you use the reaction in this section to prepare several gas jars of the 'inactive gas'? Devise an apparatus to do this and explain how it would work.

Summary

Pupils should now know that when copper is heated the black coating which appears is formed at the same time as a portion of the surrounding air is used up. The experiment has also shown that air contains at least two gases, as one-fifth of the total is 'active' and the remainder is 'inactive'.

A3.2
The history of the discovery of oxygen

In section A3.1 pupils have seen that when copper is heated in air, only one-fifth of the air ('active air') is used up. Attention is now directed to an important piece of chemical history, as pupils are shown how 'active air' was discovered. For the first time the gas is named *oxygen*.

A suggested approach

Pupils should be told of the discovery of 'active air'. They can first be told that if mercury is heated for several days in contact with air,

Objectives for pupils

1. Familiarity with the story of the discovery of oxygen
2. Knowledge of the laboratory test for oxygen

Useful films

'History of discovery of oxygen'

'O for oxygen'

BBC programme 'Oxygen'

See Appendix 4 for brief descriptions and further details of films and BBC programmes

a small amount of red 'ash' is formed on its surface and, as with copper, one-fifth of the surrounding air is used up. This experiment, however, is not suitable for demonstration in school laboratories because of the poisonous nature of mercury vapour. They should then be told that Priestley, in 1774, heated some of this red 'ash' by focusing the rays of the Sun on it using a large magnifying glass, and decomposed it into mercury and 'active air'. This was the first time that 'active air' had been isolated. This discovery can be demonstrated to the class; details of a modern version of the experiment now follow.

Experiment A3.2

How Priestley first obtained 'active air'

Apparatus

The teacher will need:

4 test-tubes, 100 × 16 mm

Bung to fit test-tube carrying delivery tube

Wide shallow beaker or small trough

Stand and clamp

Bunsen burner and asbestos square

Splints

Spatula

Mercury(II) oxide

Procedure

Place a spatula measure of mercury oxide in a 100 × 16 mm test-tube which is fitted with a bung carrying a delivery tube. Clamp the test-tube so that the lower end of the delivery tube is held beneath the surface of the water in a wide shallow beaker or small trough. Fill two or three other test-tubes with water taken from the beaker and invert them in it.

Heat the mercury oxide gently with a low non-luminous Bunsen burner flame. Take care to heat only the lowest part of the test-tube. The gas evolved should be collected in the inverted test-tubes in the beaker; the gas in the first test-tube may be tested but it will contain, mainly, displaced air.

A ring of droplets of condensed mercury will be seen above the heated portion of the test-tube. When no more gas is evolved, the delivery tube should be removed from the beaker *before* heating is stopped, otherwise water will be forced into the hot test-tube.

The drops of mercury may be examined and then returned to the laboratory stock of impure mercury; they should not be heated as mercury vapour is very poisonous. The gas can be tested with a glowing splint, a method with which the pupils should already be familiar (A2.4).

When the pupils have seen this experiment, they can be told something of Lavoisier's work in this field. Priestley showed Lavoisier his experiment on heating mercury 'ash'. After this, Lavoisier prepared a quantity of the 'ash' from the metal, and then decomposed it as Priestley had done, but this time carefully measuring the weights of the solids and the volumes of air, before and after each reaction. He found that mercury absorbed one-fifth of the surrounding air when burning, and that this same volume of 'active air' was obtained when the 'ash' was decomposed. He named the 'active air' *oxygen*, and realized that when mercury burned, it removed the oxygen from the air. This meant that air must contain at least two substances; the remaining four-fifths he called *azote*; it is now known as *nitrogen*.

In the work that follows, pupils should now use the names oxygen and nitrogen for the two major components of air. They can be reminded of the properties of the gas obtained by heating potassium permanganate, and will thus realize that this compound evolves oxygen on heating.

Finally, in this Topic, if time allows, teachers may like to tell their pupils about the modern industrial preparation of oxygen and nitrogen by the liquefaction of air, and to discuss some of the uses of these gases. Film loop 1–6 'Liquid air fractionation' can be shown here, and reference should be made to Alternative B, section B3.3, for notes concerning its use.

Suggestions for homework

1. Read the Study sheet *Burning and Lavoisier*.
2. Find out more about the life of Lavoisier or Priestley, from the school or local library. Why are their experiments so important?
3. Make a short summary of your work in this Topic.

Summary

In this Topic the pupils have seen that when copper is heated in air, one-fifth of the volume of the air is used up – the 'active' part of the air. They are told that mercury behaves similarly and are shown that this active part can be got back from the mercury ash by heating.

At this point 'active' air is named oxygen, and shown to support combustion exceedingly well, having the ability to rekindle a glowing splint.

Topic **A4**

The problem of burning

Purpose of the Topic

The purpose of this Topic is to explain the role of oxygen in combustion, and develop a classification of oxides as acidic, alkaline, and neutral.

Contents

A4.1 What are the properties of the compounds formed in burning?
A4.2 Do other substances that burn use oxygen?
A4.3 Investigating changes in weight when a candle burns

Timing

Allow a double period for each section. A maximum of four double periods for the complete Topic will give time for any extra discussion or experiment necessary.

Introduction to the Topic

In this Topic the emphasis is placed on oxygen and its function in burning. By the end of the Topic the pupils will have seen that substances use up oxygen when they burn and that they gain in weight on burning – or rather that the products of burning weigh more than the substance burnt. They will also have seen that the products of burning, the oxides, can be conveniently classified in three groups:

1. Those which react with water to form solutions which are acidic.
2. Those which react with water to form solutions which are alkaline.
3. Those which do not dissolve in water.

Background knowledge

Pupils come to this Topic with the knowledge that copper and mercury when heated in air absorb one-fifth of its volume, and that this fraction is known as oxygen. They know that oxygen can be obtained by heating mercury oxide or potassium permanganate, and that splints burn brightly in the gas.

Subsequent development

Combustion and oxides are considered again in Topic A6 'Competition among the elements'. Water as a product of combustion is investigated in detail in Topic A7, and a knowledge of some of the properties of the oxides is required in the investigations of Topic A9.

Further references

for the teacher

This Topic is closely linked to Topic A3. Additional experiments on the same theme are to be found in *Collected experiments*, Chapter 3 'Air and combustion'.

Supplementary materials

Films
'Combustion'
'Fire chemistry'
'The air, my enemy'
BBC programme 'Burning'
See Appendix 4 for a brief description and further details of films and BBC programmes

Reading material

for the pupil

Study sheet:
Fresh air? deals with some topical problems to do with the air, such as pollution, in a form simple enough to allow pupils to discuss the questions involved and to make them feel at ease in dealing with scientific concepts.

A4.1
What are the properties of the compounds formed in burning?

In this section a number of oxides are made and examined. The exercise will extend the pupils' experience of chemical reactions and give them an exercise in categorizing compounds according to their properties. It will lead up to the important consideration of elements in Topic A5.

A suggested approach

Objectives for pupils

1. Ability to classify oxides as acidic, neutral, or alkaline on the basis of pH tests
2. Awareness of the classification system: metals and non-metals

By now the pupils should understand the fundamentals of burning and be ready to extend their experience of combustion by burning a number of substances in oxygen. Before letting the pupils start, make sure that they are clear about the object of the experiment. It is to burn a number of substances in oxygen, observe what happens, and test the products of combustion with Full-range Indicator.

Experiment A4.1

Apparatus

Each pupil (or pair) will need:

Experiment sheet 12

4 hard-glass test-tubes (150 × 25 mm) and corks, and test-tube rack

Small trough

Tongs

Combustion spoon

Bunsen burner and asbestos square

Teat pipette

Stick or small piece of wood charcoal

Steel wool

Magnesium ribbon (about 3 cm)

Full-range Indicator solution

Powdered sulphur

(Continued)

Burning substances in oxygen

Note. If an oxygen cylinder is not available, or the teacher prefers to let each pupil prepare his own supply of oxygen, a convenient method is to add 20-volume hydrogen peroxide solution to granules of manganese dioxide, using the apparatus shown in the diagram. The manganese dioxide catalyses the decomposition of the hydrogen peroxide to water and oxygen, and this affords a safe and steady supply of the gas. Alternatively prepare a supply of 150 × 25 mm test-tubes filled with oxygen and closed with a cork, before the lesson.

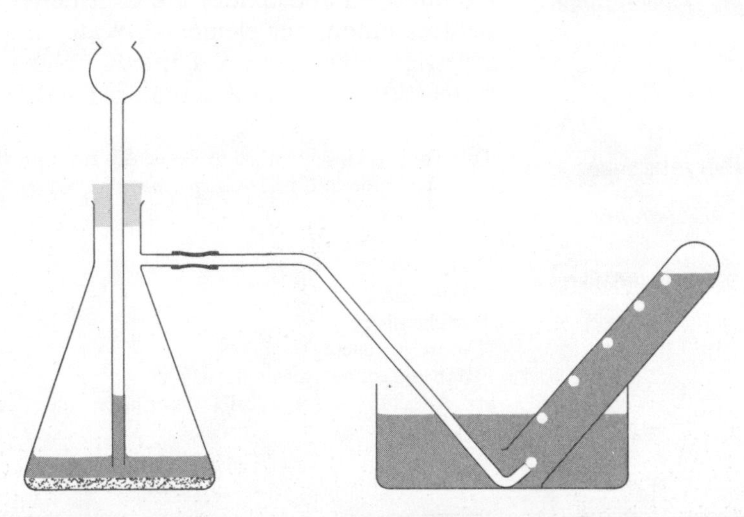

Limewater

The teacher will need:

Length of rubber tubing to lead oxygen from cylinder to pupil's beaker or trough

Cylinder of oxygen

Procedure

If the oxygen is being supplied from a cylinder, tell the pupils to fill the test-tubes with water and invert them over water in a small trough. Plastic sandwich boxes make convenient troughs. They then bring their troughs and test-tubes to the oxygen cylinder to have the test-tubes filled with oxygen. On return to their places they remove the test-tubes from the water, quickly closing them with corks, and place them in a rack ready for use. *Experiment sheet* 12 is reproduced on the next page.

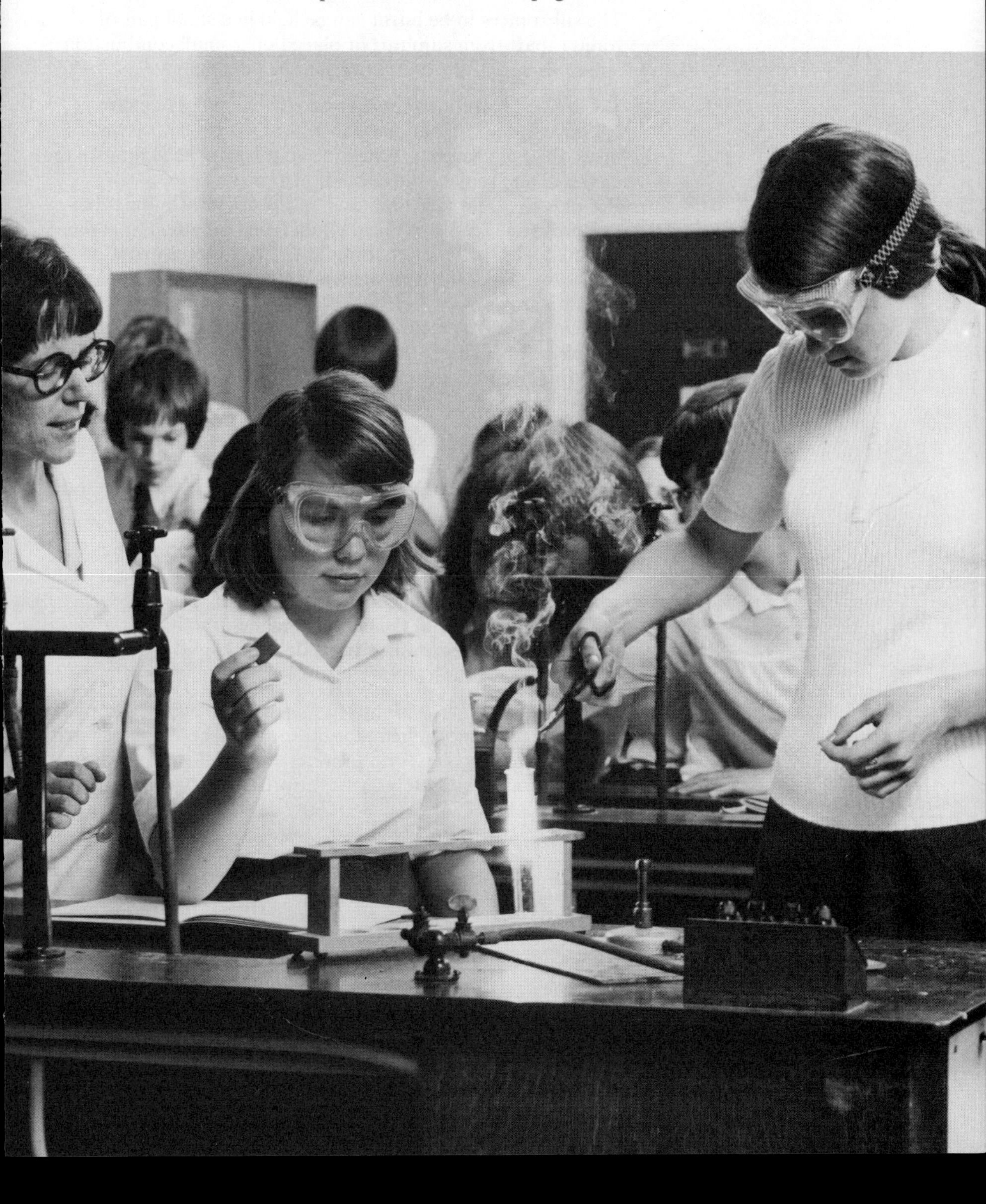

Experiment sheet 12

Many oxides can be made by burning substances in oxygen. In this experiment four oxides are made in this way and some of their properties studied.

Collect four large test-tubes full of oxygen. When you have filled each tube, cork it securely, and put it in a test-tube rack.

The substances to be burnt can be held in a small pair of tongs (apart from sulphur) or placed on a small combustion spoon. Enter your observations in the table below.

1. *Carbon.* Heat to redness one corner of a piece of charcoal. Remove the cork from one test-tube of oxygen and at once insert the hot charcoal. When the action is finished remove the charcoal and replace the cork in the tube.
Perform the following two tests on the gas now in the tube.
a. Remove a sample of the gas with a teat pipette. Expel the gas through a little water containing two to three drops of Full-range Indicator in a small test-tube. Record the colour of the indicator.
b. Remove another sample of the gas with the teat pipette and bubble it through a little clear limewater in another small test-tube. Record any change in the limewater. (This is an important test; you will often need to use it later.)

2. *Iron.* Use a small piece of steel wool or some iron powder. Heat it until it is red hot and plunge it into a test-tube of oxygen. Shake the product of burning with a little water and add a few drops of Full-range Indicator solution.

3. *Magnesium.* Use about 5 cm of magnesium ribbon held in tongs or wound round a combustion spoon. Set one end of the ribbon alight and plunge the burning metal into oxygen.
Do not look directly at the burning magnesium; it can cause damage to the eyes. Shake the product with a little water and add Full-range Indicator solution.

4. *Sulphur.* Light the sulphur first, in a combustion spoon or picked up on a moistened pipe cleaner or asbestos paper strip. Plunge it into oxygen. Test product with Full-range Indicator and limewater separately as for the product from carbon in (1).

Substances burnt	Observation during burning	Name of product	Appearance of product	Indicator colour	Action of limewater
Carbon					
Iron					
Magnesium					
Sulphur					

After the experiment discuss the results with the class. The results may be correlated by drawing a pH scale on the board and marking the values indicated for the various solutions. Ask the pupils what they think is the significance of these results. Can they see any difference between those substances which when burnt give acidic solutions, and those which give alkaline or neutral solutions? This is an opportunity to introduce the concept of metals and non-metals. According to Lavoisier's theory, the oxides of non-metals produce acidic solutions in water. Some metal oxides produce an alkaline solution in water. Many metal oxides are, of course, insoluble. It is not necessary to go into acid-base theory at this stage as it will be studied more thoroughly later on (see Stage II).

Note that the word element need not be used here; indeed it should not be, as pupils do not yet know what it means. This idea will follow in Topic A5.

Suggestion for homework

Write out a list of all the substances that you think are metals (or metallic) and those which are non-metals (or non-metallic). What properties do you associate with each group?

Summary

Pupils in this section should have acquired some additional skill in handling gases, and should know that carbon, iron, magnesium, and sulphur all burn to form oxides. They should be able to classify these oxides into acidic, alkaline, and neutral, and relate acidic oxides to non-metals, and alkaline and neutral oxides to metals.

A4.2 Do other substances that burn use oxygen?

Pupils now know that when copper, mercury, and some other materials burn they use up oxygen present in the air. In this section they try to find out whether oxygen is also used up when a candle burns. They then identify the products of combustion, and thus find out that candles must contain carbon.

A suggested approach

Objectives for pupils

1. Further recognition that oxygen is necessary for combustion
2. Understanding of the meaning of the words combustion and products of combustion

Open the lesson with a discussion of the evidence which has been gathered so far for the part that oxygen plays in burning. The pupils have seen that oxygen is used up when copper is heated in air and that carbon, sulphur, magnesium, and iron react with oxygen to form oxides. This explains why the copper gains in weight and it may also explain why some other metals gain in weight when they are heated. But what if other substances are heated, or burn, in air? Ask them to suggest how this could be tested in the case of a candle. Introduce the word 'combustion' and the phrase 'products of combustion'. A very simple experiment with a candle and a beaker is described on the next page.

Experiment A4.2a

What happens when a candle burns in air?

Apparatus

Each pupil (or pair) will need:

Experiment sheet 13

Short candle, about 5 cm long

Large beaker, approximately 400 cm^3, with no spout, *or* jam jar

Watch-glass

Full-range Indicator paper

Procedure

Burn a candle on a watch-glass in a confined volume of air under a large beaker, placed on the bench. The candle soon begins to go out. Before it does so, admit more air by raising the beaker slightly at one side. Note the mist which forms on the inside of the beaker.

Experiment sheet 13 is reproduced below.

Experiment sheet 13

Fix a candle to the middle of a clock-glass placed on the bench. Light the candle and place a large beaker or jam jar over it. What happens?

Repeat the experiment but try to keep the candle alight by raising the beaker slightly at one side when the flame begins to die down. See if you can keep it alight for five minutes. What happens to the inside of the beaker?

What do you think the products of combustion might be?

Plan a method of testing your guess. Describe it and state what happens when you try it out.

Test the inside of the beaker with Full-range Indicator paper. What happens?

Now discuss the results with the class. Why did the candle go out? Ask the class what they think the mist might be. It looks like water. Ask them how they might collect more of it, so that they could test their theory. Ask them if they think anything else might have been produced (remind them of the type of substances that are produced when sulphur and carbon burn). Suggest testing with Full-range Indicator and limewater. Lead the pupils to the idea of drawing the products of combustion over anhydrous copper sulphate (first) and then through Full-range Indicator, and lastly through limewater.

Experiment A4.2b

The products of combustion of a candle

Apparatus

The teacher will need:

Thistle funnel with bent stem

(Continued)

Procedure

Clamp the apparatus in the positions shown in the diagram and connect to the filter pump. Light the candle and wait for the drops of water to condense onto the anhydrous copper sulphate, and the

3 test-tubes (150 × 25 mm) with two-holed bungs and delivery tubes to fit

Filter pump

Stand and 2 clamps

Watch-glass and candle

carbon dioxide to turn the Full-range Indicator orange and the limewater milky.

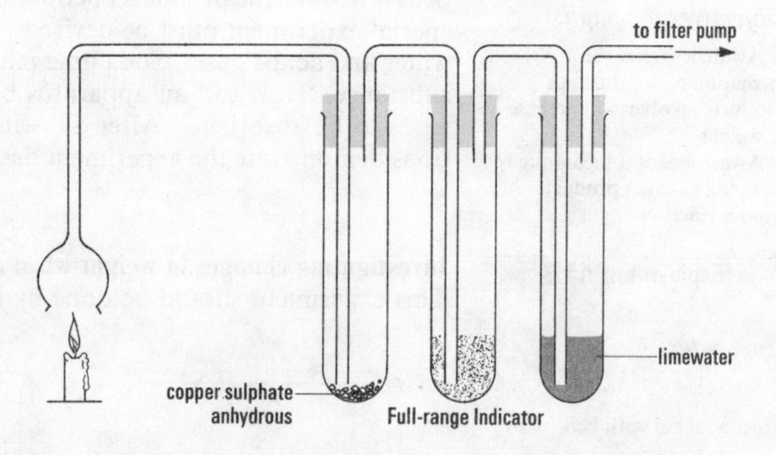

Discuss the results of the experiment with the class. The products of combustion are water and carbon dioxide. Can they suggest what the candle might contain? They might suggest carbon. Leave the discussion about the water until Topic 7. The experiment can be repeated with another substance which burns easily, such as the laboratory gas supply, and the results compared with those for the candle.

New words

Combustion
Products of combustion

Suggestions for homework

1. Write a list of all the things which you have seen burn and describe their burning. (Do **not** do any experiments on burning at home!)
2. How are fires put out? Why do you think water and sand are used?
3. Carbon dioxide is used to put out small petrol fires. Why do you think water is not used to extinguish such fires?

Summary

In this section it has been seen that other substances need oxygen to burn; and that candles, for example, form carbon dioxide and water on combustion. Pupils should now know that the word combustion means burning, and that when substances burn in air, they are combining with oxygen to form oxides.

A4.3 Investigating changes in weight when a candle burns

The final step in this Topic concerns the apparent loss in weight when things burn. An experiment is devised to find out what happens if the products of combustion are taken into account. As a result the pupils will find that there is actually a gain in weight.

A suggested approach

Objectives for pupils

1. Awareness that the formation of combustion products involves an increase in weight
2. Awareness of a technique for trapping gaseous products from a reaction

Ask the pupils how they think a candle changes in weight when it burns. They will rightly answer that it loses weight. 'But what happens if we take the products of combustion into account?' For this a special experiment must be devised using a substance that absorbs water and acidic gases. Soda lime may be introduced as just such a substance. 'How can an apparatus be made which will enable the gases to be absorbed?' After allowing the pupils to make suggestions demonstrate the experiment described below.

Experiment A4.3

Investigating changes in weight when a candle burns

This experiment should be done by the teacher.

Apparatus

The teacher will need:

Thistle funnel with bent stem

U-tube: 10 cm limbs fitted with bungs and delivery tube

Filter pump

Stand and clamp

Watch-glass

Candle

Glass-wool and tongs or forceps

Access to balance

Soda lime: granular form

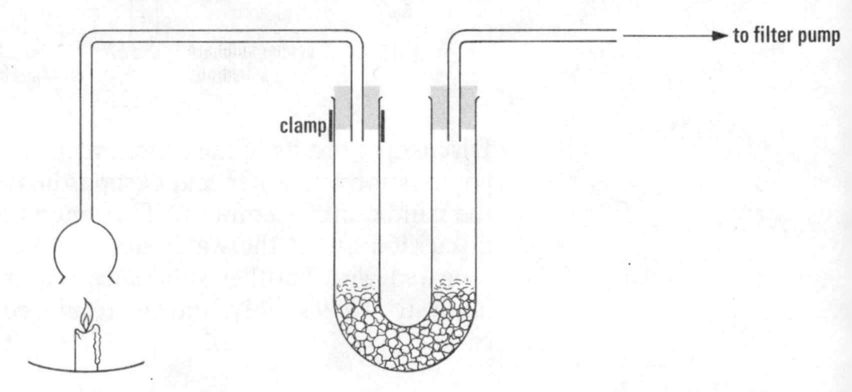

Procedure

The U-tube is loosely packed with soda lime granules to a depth of about 5 cm in each limb; the granules are held in place by small tufts of glass-wool. Fit the thistle funnel and delivery tube and weigh the entire apparatus complete with the candle and watch-glass.

Support the apparatus by a clamp and connect it to the filter pump. Light the candle and turn on the filter pump. Allow the candle to burn steadily for two or three minutes and then extinguish it and at the same time disconnect the filter pump. Do not turn the pump off. When the apparatus has cooled, reweigh it together with the candle and watch-glass.

Finally, repeat the experiment without lighting the candle but maintaining the same rate of flow by reconnecting the filter pump for the same length of time.

This increase in weight will indicate the amount of moisture and carbon dioxide absorbed from the atmosphere during the combustion of the candle. The total weight will probably increase by at least 0.3 g in three minutes.

After the experiment discuss the result with the class. They have now seen that the products of combustion of candle wax weigh more than the candle wax used up; the same thing happened when copper was heated in air.

Suggestions for homework

1. Read the Study sheet *Fresh air?*
2. Examine a candle very closely. Watch it burn and answer the following questions:
a. Why does the wick always stay about the same length?
b. In the olden days people used trimmers to cut the wick – why is it no longer necessary?
c. How does the wax get up the wick?
d. Why does 'blowing' put the candle out?
(Many other similar questions may be devised.)
3. Make a summary of your work in this Topic.

Summary

In this section pupils will have seen that there is an increase in weight when a candle burns, provided that arrangements are made to catch all the products of combustion. They will also have learned another technique used in trapping gases that are evolved from a chemical reaction.

Topic A5

The elements

Purposes of the Topic

This Topic seeks to create an awareness of what a chemist means by the term 'element', and how he decides whether or not a substance is an element. It also deals with some methods of classifying elements.

Section A5.2 is relevant to the general aim (2) (see page 13), that pupils should 'Seek patterns in the behaviour of substances'. Chapter 3 of the *Handbook for teachers* has a section on elements and the Periodic Table, pages 56–59.

Contents

Timing

The first section A5.1 will probably require more than one double period by the time the necessary experiments and discussion have been completed. One double period should be sufficient for A5.2. Thus three double periods should be the maximum time given to this Topic. This is an important Topic and deserves at least two double periods.

Introduction to the Topic

The Topic begins with the introduction of the idea of an element as a chemical which cannot be broken down into any other substances. Once this has been established the properties of some elements are examined so as to find out if the elements themselves can be classified. This is the first attempt to arrange substances in categories and the idea needs to be discussed in full. There are many properties which may be used for this purpose; colour, hardness, melting point, and chemical properties are examples. The criterion for choosing a particular property as a basis for dividing the elements into groups must be the usefulness of the resulting classification. We must ask in each case: 'Is this a *useful* way of classifying the elements?'

By the end of this Topic the pupils should know what a chemist means by an element and how he decides whether a substance is an element or not. They should also have some understanding of the way in which it is possible to classify the elements into such groups as metals and non-metals.

Alternative approach

An alternative approach to the elements is given in Alternative B, Topic B7. No practical work is included, nor is any attempt made formally to classify elements in this alternative; and the discussion depends upon the Topic on electrolysis being covered first.

Background knowledge

Pupils already understand what is meant by a single substance, and have some idea of the difference between a mixture and a 'compound' largely from the different methods of separation required;

distillation, chromatography, and so on, for separating mixtures, and electricity or heat for separating compounds. This knowledge must now be built on to establish the idea of elements as the ultimate product of all separation techniques.

Although pupils have not heard the word 'element' so far, they have met the following elements in earlier sections of Alternative A:

Element	Section
Carbon	4.1
Copper	1.6, 2.1, 2.3, 2.5, 3.1
Gold	1.6
Iodine	2.1
Iron	4.1
Magnesium	2.1, 2.3, 4.1
Mercury	3.2
Oxygen	3.2, 4.1
Sulphur	4.1

Subsequent development

The theme of elements and their properties is continued in Topic A6 'Competition among the elements'.

Reading material *for the pupil*

Study sheet:
The chemical elements. The theme of this is how the idea of the elements has developed in history with the discoveries by such famous scientists as Boyle, Davy, Mendeleev, and Ramsay. In describing this development, the Study sheet aims to help pupils acquire the idea of what an element is.

A5.1 What is an element?

In the course of the first four Topics the pupils have learnt what chemists mean by 'a pure substance' and have met a number of chemical reactions, including some involving decomposition. In the last Topic metals and non-metals were discussed. The purpose of this section is to introduce the concept of elements.

A suggested approach

Objectives for pupils
1. Development of the chemists' concept of an element
2. Understanding of the meaning of the word element

Remind the pupils that they started the course with a series of experiments which led them to the concept of a pure substance. To them it is something which cannot be separated into two different substances by techniques such as filtration, distillation, crystallization, or paper chromatography. They obtained pure salt from rock salt by crystallization. No matter how many times they re-crystallized the salt they could not divide it up into any component parts. They separated the green stuff from grass and the black stuff from ink. They found that the green and black materials could be further split up by paper chromatographic methods. But once this separation had been made no amount of paper chromatography could split up the substances obtained any further. In each case a pure substance had been isolated. They know then, in general terms, what a pure substance is. But they also know that some of these pure substances – crystalline copper sulphate is an example – can be broken down into two or more substances by other, usually more drastic, means. Heating makes the apparently 'pure' crystals of blue hydrated copper sulphate break down into white anhydrous copper sulphate and water. Electrolysis breaks copper sulphate down yet further and copper is obtained. But copper, however strongly it is heated in the absence of air, does not show any tendency to break down. Thus there does seem to be an end to all this breaking down; a stage is reached at which it is no longer possible to divide any particular substance into other substances by either heat or electricity.

The idea of all things being made up from other things is ancient. The basic or elemental things from which other things were thought to be made up were called elements. In Aristotelian philosophy these elements were earth, air, fire, and water, and an extensive 'chemistry' was built up on this theory.

It was Robert Boyle (1627–1691) who suggested that an element should be thought of as a substance which could not be broken down into any further substances. By this definition there is no absolute way of determining whether a substance is an element or not. We can only say that nothing we do to this substance breaks it down any further; we therefore assume that it is an element. Tell the pupils, if you wish, that since the discovery of mass spectrometry the definition of the term 'element' has been modified and that it is now possible to determine with certainty whether a given substance is an element or not.

Having established the idea of elements, take a selection of elements and have a look at their physical properties. A suitable selection would be carbon, lead, copper, aluminium, and sulphur. A series of experiments to test the properties of these elements is given below. These experiments demonstrate the physical differences between metals and non-metals.

Experiment A5.1a

An investigation of some elements

Apparatus

Each pupil (or pair) will need:

Experiment sheet 14

Bunsen burner and asbestos square

6 V bulb and bulb holder

6 V battery or alternative d.c. supply

Connecting wires and crocodile clips

Hammer

Access to balance

Procedure

Details are given in *Experiment sheet* 14 reproduced below.

Experiment sheet 14

So far you have met a number of substances which are made up of two or more other substances, such as black copper oxide which is made up of copper and oxygen. Substances like copper and oxygen, which are simple substances in that they are not made up of anything but themselves, are called *elements.*

Have a good look at the specimens of some elements which have been provided for you. Test them as follows.

1. Try to decide a rough order of densities (weights of equal volumes). Weighing pieces of different elements of about equal size could help.

Carbon, sulphur, copper, aluminium, lead, and other metal rods as available, about 15 cm long and of similar diameter. *Or*, pieces of metal foil about 6 cm × 2 cm, of about the same thickness.

Which seems to be the densest element?
Which is the denser – copper or aluminium?
Which seems to be the element of lowest density?

2. Test small pieces of the solid elements by tapping them with a hammer or a similar solid object.
Which of them are brittle?

3. Compare the effectiveness of the elements as conductors of heat by heating one end of wires or rods made of them, while holding the other end in your hand. (**Do not** try this for sulphur.)
Which of them conduct heat well?
Which is the best conductor?

4. Test each element with a battery and torch bulb to find whether or not it conducts electricity.
Which elements conduct electricity well?
Which are poor conductors?

After discussing the results of testing the above elements for their physical properties, group them on the blackboard according to whether they are (*a*) brittle or ductile, (*b*) good or bad conductors of heat and electricity, and (*c*) more or less dense. Now suggest testing the chemical properties of some more elements. This may be done by burning them in air; lead, zinc, calcium, and sodium are suitable for this purpose.

Experiment A5.1b More properties of elements

Apparatus

The teacher will need:

Crucible lid or piece of broken porcelain

Asbestos paper strip, approximately 7 cm × 2 cm

Tongs

Steel knitting needle or iron nail

Bunsen burner and asbestos square

4 test-tubes, 100 × 16 mm

Small quantities of lead foil, zinc dust, calcium turnings, and sodium

Full-range Indicator solution

Filter paper

Petroleum spirit

Methylated spirit

Notes

1. Because of the possibility of accidents when using sodium, this experiment should be demonstrated by the teacher. Sodium should be handled in the following manner.

Store small pieces of the metal in liquid paraffin and remove them with a pair of tongs. The pieces of sodium must then be cleaned by dropping them into a beaker containing petroleum spirit. Remove the pieces from the beaker and allow them to dry on filter paper. They may be cut with a penknife but must not be touched or allowed to come into contact with any water. Place and leave any small pieces not used in a beaker containing methylated spirit until effervescence stops; the liquid may then be safely thrown away.

2. It is advisable to restrict the use of lead and its compounds, and demonstrating this experiment will achieve this object also.

Procedure

The object is to 'widen the front' and heat more metals in the air.

1. Heat a piece of sodium the size of a rice grain on a crucible lid or piece of broken porcelain. Note the ease with which it burns and the colour of the flame.

2. Small pieces of lead foil should be heated on a crucible lid (or

piece of broken porcelain), held with tongs, for several minutes. Stir the molten globule often with a steel knitting needle to expose the metal to the air. A yellow coloration is observed; on cooling, examine the residual ash.

3. Place a spatula measure of zinc dust near the edge of a strip of asbestos paper and heat it. As soon as the zinc catches fire and burns, take the paper out of the flame and gently prise up the yellow crust that forms with a steel knitting needle. The metal, now exposed to the air, burns, forming the white fluff of 'Philosopher's Wool'.

4. Hold a piece of calcium (usually supplied in the form of turnings) by the tongs and heat it strongly in a roaring Bunsen flame. It is not easy to ignite because the tongs conduct the heat away and make it difficult to raise the temperature of the calcium to its ignition point. However, by using asbestos paper to sandwich the calcium turning, this difficulty can be avoided. Note the colour of the flame and the nature of the product.

Examine the action of water on the 'ashes' formed from all these metals, and test the liquid with Full-range Indicator solution.

Discuss the results of this experiment with the class and compare the results with those of Experiment A4.1.

New word

Element

Suggestions for homework

1. Read the Study sheet *The chemical elements.*
2. Make a list of the various substances in your kitchen. How many of them are elements?
3. How many substances do you know that are elements? What makes you think that they are elements?

Summary

Pupils should now know the meaning of the term 'element', and something of the properties of several common elements, including their densities, brittleness, conductivities of heat and electricity, and, in the case of metals, how they burn in air.

A5.2
Into what groups can we sort elements?

In this section the idea that elements can be divided into groups, other than just metals and non-metals, is introduced. It is not intended that the groups of the Periodic Table should be mentioned at this stage, but merely that some small groups of similar elements, such as magnesium, calcium, and sodium, should be recognized.

A suggested approach

First discuss the results of the questions set for homework. How many elements did they identify? How many others do they know? It might be worth writing them on the board. They are oxygen,

Objectives for pupils

1. Knowledge of some methods of classifying elements
2. Ability to use a table of data

nitrogen, carbon, sulphur, iron, magnesium, copper, lead, zinc, calcium, sodium, iodine, and aluminium (if some of the experiments have been omitted from your course, leave the relevant elements out). The pupils should be able to divide these into metals and non-metals using the knowledge they have gained so far from experiments. One way to do this is to let the pupils write them out in two columns on a piece of paper. Then ask them to say which elements they have put in each column and to explain why they have put them there.

The next question is: 'Into what groups can the elements be further divided?' Of course they are unlikely to be able to do this without help at this stage, but with help they can divide the elements into simple groups. Ask the class to look at the metals. Can they see any similarities or differences between them? What about their physical properties?

Are they hard or soft? Are they 'silvery' (white metals) or not? Do they differ in chemical properties? What happened when they were burnt and their oxides were dissolved in water? They will see that with the metals they have studied so far it is the chemical differences which divide them most effectively. On the one hand there are metals which produce oxides which give alkaline solutions with water and on the other hand there are those whose oxides do not appear to dissolve at all. Allow the pupils to test samples of oxides that they have either not tried for alkalinity or whose effects they have forgotten. This may be done by simply shaking a little of the oxide in a test-tube with about 5 cm^3 of distilled water, and then testing with Full-range Indicator paper or solution. They will find that, of the metals, only magnesium, calcium, and sodium give an alkaline solution. These could be put together as a special group of metals.

Suggestions for homework

1. Prepare an abridged list of elements and their densities, suitable for the class you are teaching, and ask the pupils to divide the elements into:
a. Non-metals.
b. Metals with high densities (over 5 g cm^{-3}).
c. Metals with low densities (under 5 g cm^{-3}).
A suitable list might be as follows.

The densities of some elements in g cm^{-3}

Element	Density	Element	Density
Aluminium	2.70	Phosphorus (white)	1.83
Calcium	1.54	Platinum	21.37
Carbon (graphite)	2.22	Potassium	0.87
Copper	8.94	Silicon	2.42
Gold	19.32	Silver	10.50
Iodine	4.94	Sodium	0.97
Iron	7.86	Sulphur	2.07
Lead	11.35	Tin	7.31
Lithium	0.54	Tungsten	19.10
Magnesium	1.74	Uranium	18.68
Mercury	13.55	Zinc	7.13

2. Make a short summary of the work you have done in this Topic.

Summary

Pupils should now be able to classify elements into metals and non-metals, and be able to see other smaller groupings within these broad divisions.

Topic A6

Competition among the elements

Purposes of the Topic

In this Topic it is intended that pupils should become familiar with the idea that elements can be put in an order of reactivity, and that from this they can predict the likelihood of reactions taking place.

This Topic introduces the third of the general aims of the course, that pupils should 'Develop concepts related to the classification of substances and their observable behaviour', and it is hoped that pupils will be excited by the possibility of *predicting* the course of reactions, and testing their predictions. This important theme occurs throughout the Nuffield chemistry courses and is still being considered in the last Topic of the Advanced Chemistry course!

Contents

Timing

The time taken for this Topic is likely to vary a great deal. Four double periods should be the maximum time to spend. Cut to a minimum, this Topic could be covered in two double periods.

Introduction to the Topic

This Topic is concerned with the idea that elements can be arranged in the order of their reactivity with another element. So many factors influence the apparent reactivity of an element that the subject is approached with caution. In the first section reactivity with oxygen is discussed and an 'order of reactivity' is drawn up. The order is then tested in terms of a 'competition for oxygen' between magnesium and a metal oxide. In the second section experiments are performed to see where carbon fits in this series.

Alternative approach

The subject is treated rather differently in Alternative B, Topic B10.

Background knowledge

This Topic follows naturally from the consideration of burning in Topics A3 and A4 and from the idea of elements in Topic A5.

Subsequent development

The ideas developed in this Topic are used again when dealing with the extraction of iron from iron ore (Topic A9) and in comparing the reactivities of the halogens (Topic A10).

Further references
for the teacher

Additional experiments on this theme are to be found in *Collected experiments*, Chapter 7 'Reactivity series'.

Reading material
for the pupil

Study sheet:
Competitions. The word 'competition' is one of several whose meaning in chemistry differs widely from its other meanings in common usage. Pupils frequently find it difficult to distinguish these usages at this stage, and the Study sheet aims to help them over this.

A6.1
Putting elements in order of their reactivity with oxygen

In this section the pupils' experience of burning elements in the air is used to build up a reactivity series. It is stressed that this only tells us directly about the reactivity of the elements with oxygen and that many other factors affect the apparent reactivity of elements. The reactivity series is then used to predict whether magnesium oxide will react with copper or copper oxide with magnesium.

A suggested approach

Objectives for pupils

1. Recognition that there is a pattern in the vigour of reactions
2. Awareness that an order of reactivity enables reactions to be predicted
3. Knowledge of the order of reactivity of some common metals
4. Understanding of the meaning of the words reduce and reactivity

The emphasis up to now has been on what changes are taking place in a chemical reaction. The question of how vigorously reactions take place has not yet been raised although the pupils are bound to have noticed it, particularly when they heated substances in air (A4.1). This point may be used to open a discussion at the beginning of the first lesson in this Topic. Ask the pupils for the names of metallic elements that they remember heating or burning and write the names on the board. Examples are iron, magnesium, lead, zinc, sodium, copper, and calcium. Which of those do they think reacted most vigorously with air? 'Does copper burn in air?' 'Does iron-wool burn as brightly as magnesium?' With some discussion and help a list may be made of the elements in order of their apparent reactivity. It should be stressed that this is only their *apparent* reactivity, as many different factors affect the vigour with which the elements react. State of division is one such factor, which may easily be demonstrated by heating a little zinc powder on a strip of asbestos paper and then heating a solid lump of zinc. A rough grouping based on this discussion would be (*a*) sodium, magnesium, and calcium, (*b*) zinc, (*c*) iron, (*d*) lead and copper. Once the list is on the board it can be used. Ask for suggestions about the reactions which it could predict. 'How can you remove the oxygen from copper oxide?' 'Where does copper appear in the list of elements?' 'Will iron react with calcium oxide?' 'Why?' According to the list magnesium is more reactive than copper. 'Does this mean that magnesium will take the oxygen away from copper oxide?' The question can be answered by trying it; details of the procedure, *which should be followed closely*, are given below.

Experiment A6.1a

The reactivity of magnesium and copper
This experiment **must** be demonstrated by the teacher.

Apparatus

The teacher will need:

Plastic safety screen

Small crucible or piece of asbestos paper

Tripod

Pipe-clay triangle (if crucible is used)

(Continued)

Procedure

Heat about one spatula measure of a mixture of equal parts of magnesium powder and thoroughly dried copper oxide – or lead oxide – in an open crucible or on a piece of asbestos paper. After a few moments a violent reaction occurs. In view of this, the teacher should place the Bunsen burner under the mixture and stand back until the reaction is over.

The pupils will expect to see some copper at the end! Very little remains, so repeat the experiment but moderate the reaction by

Bunsen burner and asbestos square

Beaker, 100 cm^3

Filter funnel and paper

Stand and clamp

Spatula

2M hydrochloric acid

Magnesium powder

Copper(II) oxide or lead(II) oxide (litharge)

mixing in some magnesium oxide. Then, either dissolve away the magnesium oxide at the end with 2M hydrochloric acid and filter off the metal residue, or do a control experiment with magnesium powder only. The reaction with copper oxide is seen to be much more vigorous and the residue is very different.

Experiment A6.1b

Apparatus

The teacher will need:

Plastic safety screen

Silica crucible

Tin of sand

Asbestos squares

Spatula

Taper

Aluminium powder

Red iron(III) oxide (3 to 5 g is sufficient and is thoroughly dried by heating in an evaporating dish over a Bunsen burner)

Barium peroxide

Magnesium powder and ribbon

The reactivity of aluminium and iron

This experiment **must** be done by the teacher.

Procedure

Mix together about equal volumes of dry aluminium powder and thoroughly dried red iron oxide (see list). Place the mixture in a silica crucible and stand this in a tin filled with sand. Place a spatula full of a mixture of barium peroxide and magnesium powder on top of the first mixture and insert a piece of freshly scraped magnesium ribbon through this pile to act as a fuse. Light the magnesium ribbon with a taper fixed into the end of a long glass tube and stand well back. *An extremely vigorous reaction takes place*, quantities of light and smoke are emitted, and a bead of iron can be emptied out when the reaction has subsided.

This is known as the thermit reaction.

After this experiment let the pupils try a few examples for themselves. *Do not allow them to heat any of the oxides with sodium, magnesium, or calcium as these reactions can be dangerous.*

An experiment for pupils is described below. During the work the word *reduce* – meaning (at this stage) to remove oxygen from a substance – may be introduced. (See *Handbook for teachers*, Chapter 2, pages 31–34.)

Experiment A6.1c

Apparatus

Each pupil (or pair) will need:

Experiment sheet 15

Asbestos paper strips, approximately 7 cm × 2 cm

Bunsen burner and asbestos square

Tongs

Iron powder

(Continued)

Competition for oxygen between iron, zinc, copper, and lead

Procedure

Pupils should mix together a little of one of the oxides with an equal volume of iron powder, and heat the mixture on a strip of asbestos paper. If a reaction occurs a glow will spread through the mass. They should then repeat this with the other oxides.

Experiment sheet 15 is reproduced below.

Experiment sheet 15

Some elements combine with other elements to form compounds with great vigour; others seem to have to be persuaded to combine with each other. You will have seen

Experiment sheet 15 *(continued)*

Zinc oxide

Copper(II) oxide

Lead(II) oxide

already what happens when two elements are competing for a third element. You are now going to investigate some further examples for yourself.

In each case mix a measure of iron powder with a measure of metal oxide. Put the mixture on a folded strip of asbestos paper. Hold the paper in tongs and heat the mixture. Describe what happens and examine the mixture to see if the oxygen has 'changed partners', that is, if a new metal has been formed. Warming the residue with dilute hydrochloric acid may help.

Investigate the following mixtures.
1. Iron and copper oxide
2. Iron and lead oxide
3. Iron and zinc oxide

Arrange the four metals involved in this experiment as far as possible in order of their affinity for oxygen, putting the one with the greatest affinity first.

New words

Reduce
Reactivity

Suggestions for homework

1. Write word equations for the reactions you have studied in this Topic.
2. How are you able to decide whether or not a reaction takes place when a metal, such as iron, is heated with a metal oxide, such as copper oxide? How can you show that new substances have been formed during the reaction?

Summary

Pupils should now know the reactivity towards oxygen of the following elements: magnesium, aluminium, zinc, iron, copper, lead. They should realize the importance of their knowledge in predicting the likelihood of chemical reactions between one metal and the oxide of another metal.

A6.2
Where does carbon come in the series?

In the last section a simple reactivity series of metals was established and tested. The relative position of a non-metal is now investigated. The importance of this empirical knowledge in industrial metallurgy is pointed out.

A suggested approach

Objectives for pupils

1. Ability to use correctly the idea of an order of reactivity

The pupils saw in the last section that magnesium could be used to remove oxygen from metal oxides. Tell them that in practice it is often necessary to remove oxygen from metal oxides on a very large scale – to make zinc from zinc ore for example. It would be far too expensive to use magnesium and in fact coke (consisting largely of

2. Knowledge of the use of coke in industry to reduce metal oxides

carbon) is used. It would be very interesting, therefore, to see where carbon fits into the reactivity series. Which metal oxides can have their oxygen removed by carbon? They may recall Experiment A1.6a.

A series of experiments may now be devised to find the answer to this question. It is important that the pupils should try some reactions which 'do not work'. For instance before trying Experiment A6.2b the action of carbon on magnesium oxide should be tried. This is one of the substances mentioned in the first experiment.

After Experiment A6.2a the pupils will see that carbon is below magnesium and above lead. To confirm that carbon is below magnesium they can try removing the oxygen from carbon dioxide with magnesium. Experiment A6.2b describes how this may be done.

Experiment A6.2a

How does carbon compare with the metals in its affinity for oxygen? (Part 1)

Apparatus

Each pupil (or pair) will need:

Experiment sheet 16

Asbestos paper strip, approximately 7 cm × 2 cm

Bunsen burner and asbestos square

Tongs

Watch-glass

Spatula

Magnesium oxide

Lead(II) oxide (litharge)

Powdered wood charcoal

Procedure

Mix together a little lead oxide with an equal volume of powdered charcoal on a watch-glass and transfer to a piece of asbestos paper folded so that its cross-section is V-shaped. Hold the asbestos paper with a pair of tongs, and heat the mixture with a Bunsen burner.

After a short time beads of molten lead appear. A bead can be shown to have one property of lead by washing it and rubbing it on paper. Lead marks the paper. Samples of lead beads on asbestos paper may be stuck into laboratory books with Sellotape.

Repeat the experiment but this time mix a little magnesium oxide with an equal volume of powdered charcoal. Even after strong heating no reaction will be seen to have taken place.

Experiment sheet 16 for this experiment is reproduced after Experiment A6.2b.

Experiment A6.2b

How does carbon compare with the metals in its affinity for oxygen? (Part 2)

Apparatus

Each pupil (or pair) will need:

Experiment sheet 16

Gas-jar and cover or corked hard-glass test-tube

Bunsen burner and asbestos square

Tongs

Magnesium ribbon, approximately 5 cm

The teacher will need a supply of carbon dioxide

Procedure

Fill a gas-jar or test-tube with carbon dioxide and seal it carefully with a greased cover or cork. Place the gas sample near a Bunsen burner. If the test-tube is used support it in a rack during this experiment.

Hold a piece of magnesium ribbon (about 5 cm long) by means of a pair of tongs and place one end in a Bunsen burner flame. As soon as it ignites plunge it quickly into the carbon dioxide; it will continue to burn for a few moments. As soon as the magnesium has finished burning, examine the contents of the gas-jar or test-tube for black specks of carbon and for white magnesium oxide.

Experiment sheet 16 gives the procedure for both Experiment A6.2a and Experiment A6.2b.

Experiment sheet 16

So far you have studied 'competition for oxygen' between the metals. You know that carbon burns quite vigorously in oxygen. You will now try to find out how it compares with the metals in its affinity for oxygen.

1. Mix half a measure of lead oxide with an equal bulk of powdered charcoal. Heat the mixture strongly on a folded strip of asbestos paper.
What happens?
Which has the greater affinity for oxygen – carbon or lead?

Now repeat the experiment, this time mixing half a measure of magnesium oxide with an equal bulk of powdered charcoal.
What happens?
Which has the greater affinity for oxygen – carbon or magnesium?
Check your answer to this experiment by doing the next experiment.

2. Here is another 'competition for oxygen' involving carbon. Fill a gas jar with carbon dioxide. Hold 5 cm of magnesium ribbon in tongs, set it alight, and plunge it at once into the gas jar. *Avoid touching the sides of the glass jar with the burning magnesium.*
Does the magnesium go on burning?
If so, where is it getting its oxygen from?
What two substances are left in the gas jar?
Write a 'word equation' for the change that has occurred:

magnesium + carbon dioxide →

Suggestions for homework

1. Read the Study sheet *Competitions*.
2. Describe and explain the experiments you would carry out to place the metals calcium, silver, and nickel in the reactivity series.
3. Predict what you would expect to happen when (*a*) carbon is heated with zinc oxide, (*b*) magnesium is heated with lead oxide, (*c*) carbon dioxide gas is passed over red hot copper. How would you try and test your predictions?
4. Haematite is an ore of iron, being a type of iron oxide. How would you try to get iron out of this ore in the laboratory?

Summary

Pupils should now know the position of carbon in the reactivity series, its place being between zinc and iron. They should also be aware that carbon is the main reducing agent used in industry to make metals from their oxides.

Topic A7

Water as a product of burning

Purposes of the Topic

The main purpose of this Topic is to use the understanding of relative reactivity and of reduction gained in Topic 6 to elucidate the composition of water. By working in this way, pupils will discover that water is hydrogen oxide by analysis. This finding is then confirmed by synthesis, and so the Topic demonstrates the role of analysis and synthesis in determining the composition of a compound. The Topic also gives pupils a knowledge of the simple chemistry of hydrogen.

Contents

A7.1 What is the liquid condensed from a Bunsen flame?
A7.2 Is water an oxide?
A7.3 What other metals can we use to obtain the inflammable gas from water?
A7.4 A more convenient way of preparing hydrogen in the laboratory
A7.5 Can hydrogen be used to reduce metal oxides?

Timing

There is a great deal of experimental work in this Topic. At least one double period should be allowed for each section. A maximum allowance of six double periods will give time for discussion and summing up. If time is short, leave out A7.4. This preparation becomes less important if a cylinder of hydrogen is used as a source of supply. The other sections could then, perhaps, be condensed into three double periods.

Introduction to the Topic

This Topic gives the pupils an opportunity to find out about the nature of a substance, in this case water, by using some of the observations and deductions they have made in previous sections. Briefly the argument is as follows: (1) water is seen to be a product of combustion (observation); (2) when substances burn in air, they are reacting with oxygen and forming oxides (generalization from a number of observations); (3) water is therefore probably an oxide (deduction from foregoing generalization). This deduction is then followed up and experiments are devised to see if water is indeed an oxide. To find out, use is made of the ideas met in Topic A6 (competition for oxygen). The pupils look for a metal with a high affinity for oxygen that will remove it from water. A recognizable oxide is formed and therefore water must contain oxygen. The inflammable gas which is produced in this reaction is named hydrogen.

In the sections which follow the pupils explore other ways of producing hydrogen. To test the hypothesis that water is an oxide of hydrogen, the hydrogen is burnt and the product examined. It is found to be water and the hypothesis is thus confirmed.

By the end of the Topic the pupils should appreciate the reasoning behind the experiments that they have performed, and know that analysis should be followed by synthesis of the compound analysed. They should be familiar with some of the simple chemistry of water and hydrogen.

Background knowledge

Water is first studied in Topic A1, when it is obtained as a result of distilling ink and its boiling point is used to identify it. The effect which it has on anhydrous copper sulphate is seen in Topic A2. It is first noticed as a product of combustion (and therefore as a likely oxide) in Topic A4 where it is found to be formed when a candle burns, and in a Bunsen burner flame.

The possibility of removing the oxygen from oxides is first mentioned in Topic A6 and a number of such reductions are carried out. The order of reactivity dealt with there provides the background knowledge required for an attempt to remove the oxygen thought likely to be present in water.

Subsequent development

Topic A10 is concerned with sea water.

Further references

for the teacher

Additional experiments are given in *Collected experiments*, Chapter 4 'Water'.

Reading material

for the pupil

Study sheet:
Water is broad in approach, to expand pupils' horizons on this substance which they see so often both inside and outside the laboratory. It deals with how water affects the lives of people all over the world, and with ways of ensuring that we have enough water in the future.

A7.1
What is the liquid condensed from a Bunsen flame?

In this first section of the Topic, burning gas from a Bunsen burner is investigated, and water is found to be a product of combustion.

A suggested approach

Objective for pupils

Knowledge of water as a product of combustion

To start the discussion, light a Bunsen burner and place it under a beaker of cold water on a tripod and gauze. Ask the pupils what they see. A mist will be seen to form on the outside of the beaker. Ask them where they think it comes from. Some may suggest the water inside the beaker; others the air, and so on. Experiments may be devised to test each of these suggestions in a simple way. Someone may suggest that the mist comes from the flame; indeed the mist can sometimes form on the outside of an empty beaker, thus disposing of the suggestion that the water inside has anything to do with the formation of the mist.

The pupils should also be asked what they think the mist is; many will suppose that it is water, but how can this be proved? Remind the pupils of section A4.2, when they investigated the products of combustion of a candle. How did they find out if water was present then? How did they collect sufficient water to test?

By questions such as these lead the pupils to suggest an experiment such as Experiment A7.1, which should then be demonstrated.

Experiment A7.1

Is water formed in a Bunsen flame?

This experiment **must** be performed by the teacher.

Apparatus

The teacher will need:

U-tube or calcium chloride tube

Thistle funnel with stem bent as in diagram, fitted with bung for the test-tube

Test-tube with side-arm, 125 × 16 mm

Beaker, 250 cm^3

2 stands and clamps

Filter pump and connecting tubing

Glass jet and connecting tubing

Test-tube, 100 × 16 mm

Thermometer −10 to +110 °C

Bunsen burner and asbestos square

Silica gel or anhydrous calcium chloride

Anhydrous copper sulphate

Access to gas tap

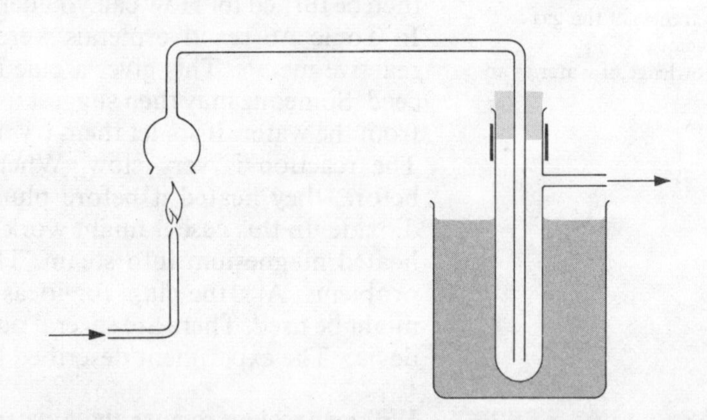

Procedure

Gas from the laboratory supply is first dried by passing it through a tube containing silica gel or anhydrous calcium chloride before the gas is burnt at a jet. The flame should be about 3 cm high and allowed to burn beneath a thistle funnel as illustrated in the diagram. A steady flow of the hot gas produced is maintained through the apparatus by the filter pump, and a liquid soon condenses in the side arm test-tube which is cooled in a beaker of water. The cooling will be even more effective if ice is added to the beaker of water. After about ten minutes 2 or 3 cm^3 of liquid will have collected. This should be sufficient to check the boiling point and examine action on anhydrous copper sulphate.

Suggestion for homework

Can you think of any evidence that water is formed when (*a*) coal, and (*b*) petrol, are burnt? What experiments could you do to find out whether water was formed in either of these cases?

Summary

Water is found to be a product of combustion, and is produced in a Bunsen burner flame.

A7.2 Is water an oxide?

The last experiment identified water as a product of the combustion of town (or North Sea) gas. All products of combustion that the pupils have met up to now have been oxides and so the question 'Is water an oxide?' is a very reasonable one. Use is made of the knowledge gained in Topic A6 'Competition among the elements' to find a way of showing that water is indeed an oxide.

A suggested approach

Objectives for pupils

1. Ability to design simple experiments
2. Awareness of the gas hydrogen
3. Knowledge of water as an oxide

The object of this section is to demonstrate that water is an oxide. All this should be discussed at the beginning of the lesson together with the results of the homework from A7.1 if this was set after the last lesson. It is very likely that one of the pupils will know that 'Water is H_2O and therefore must be an oxide'. The question can then be turned to 'How can you demonstrate that water is an oxide?' In Topic A6 reactive metals were used to take oxygen from less reactive metals. This gives a clue to the way in which we can proceed. Someone may then suggest using magnesium to get the oxygen from the water. If so, let them try magnesium ribbon on cold water. The reaction is very slow. When they used magnesium ribbon before, they heated it before plunging it into a gas-jar of carbon dioxide. In this case it might work more quickly if we could plunge heated magnesium into steam. This poses a number of technical problems. Ask the class for ideas on the sort of apparatus which might be used. There are several possibilities which the pupils might devise. The experiment described below gives one example.

Experiment A7.2

Will magnesium remove the oxygen from water?

Apparatus

The teacher will need:

Test-tube, hard-glass, 150 × 25 mm, fitted with bung and short delivery tube

Bunsen burner and asbestos square

Stand and clamp

Teat pipette

Sand paper

Taper

Asbestos paper or mineral wool

Magnesium ribbon about 15 cm long

Note

The hard-glass test-tube used in this experiment will not be recoverable, and for this reason it is suggested that teachers should demonstrate the experiment.

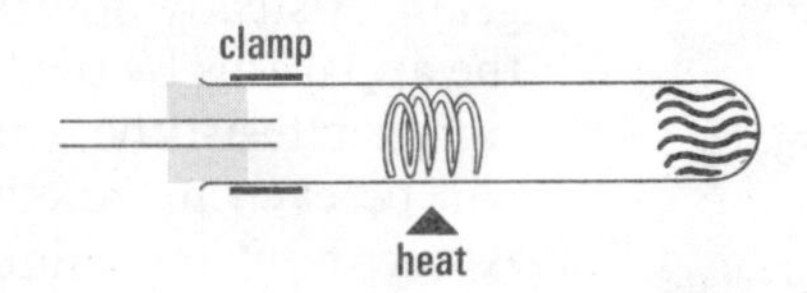

Procedure

Place sufficient loosely rolled asbestos paper in a 150 × 25 mm hard-glass test-tube to fill it to a depth of about 3 cm. Add about 2 cm^3 of water, which will be as much as the asbestos will absorb completely, and clamp the tube horizontally.

Clean about 15 cm of magnesium ribbon by means of sand paper, form the cleaned ribbon into a compact cylinder, and place it about 2 to 3 cm from the end of the wet asbestos. Fit the bung carrying the short delivery tube.

Gently heat the part of the test-tube containing the magnesium by means of a non-luminous Bunsen flame until the magnesium first appears to melt. Just touch that part of the tube containing the wet asbestos with the flame so that some steam is generated and passes over the hot magnesium. **Warning:** An intense flash occurs; the gas produced may be burnt at the end of the delivery tube when a lighted taper is applied. Continue heating the test-tube and generating the steam until the reaction is complete.

When the tube has cooled the contents may be tipped out and examined. The pupils will notice the black markings on the walls of the test-tube. If a pupil draws attention to this effect, ask whether it might be due to the reaction of magnesium with the glass.

Afterwards discuss the results. What was the white powder? It looks just like the magnesium oxide from Experiment A2.1. (Show pupils 'comparative' tests with water and indicators.) If water is an oxide, what is it an oxide of? The gas which came off was shown to be flammable. Its properties will be investigated in section A7.4.

Suggestions for homework

1. Read the Study sheet *Water*.
2. What other ways of getting hydrogen from water can you suggest?
3. Write a clear summary of the evidence which shows that water is hydrogen oxide.
4. Try to find out how hydrogen is manufactured commercially.

Summary

Magnesium is found to take the oxygen away from water, and leave an inflammable gas.

A7.3
What other metals can we use to obtain the inflammable gas from water?

In the last section an inflammable gas was obtained from water by removing the oxygen from it using magnesium. This was found to be a vigorous reaction, and in this section a more controlled reaction to produce the inflammable gas is sought. The gas is named *hydrogen*.

A suggested approach

Objectives for pupils

1. Appreciation of the use of an order of reactivity to predict reactions
2. Ability to design a suitable apparatus to produce the inflammable gas from water in a more controlled manner

As a means of producing the inflammable gas, the magnesium reaction was clearly too vigorous. Ask the pupils what other metals they would expect to react with steam less vigorously than magnesium. Going down the reactivity series they may suggest zinc or iron. The wet asbestos method may be used to try the effect of steam on the two metals.

Experiment A7.3

To investigate the action of steam on iron and zinc

Apparatus

Each pupil (or pair) will need:

Experiment sheet 17

Stand and clamp

Bunsen burner and asbestos square

Hard-glass test-tube, 125 × 16 mm, fitted with bung and delivery tube

(Continued)

Procedure

Caution. Warn the pupils about the danger of 'sucking back' before allowing them to start the experiment.

Further details are given in *Experiment sheet* 17 which is reproduced on the next page.

4 test-tubes, 150 × 25 mm, and corks

Crystallizing dish or small trough

Teat pipette

Spatula

Asbestos paper or mineral wool

Iron powder

Zinc powder

Experiment sheet 17

You have seen that steam reacts vigorously with hot magnesium. The action of steam on heated iron is likely to be less violent.
Why?

Use the apparatus shown in the diagram.

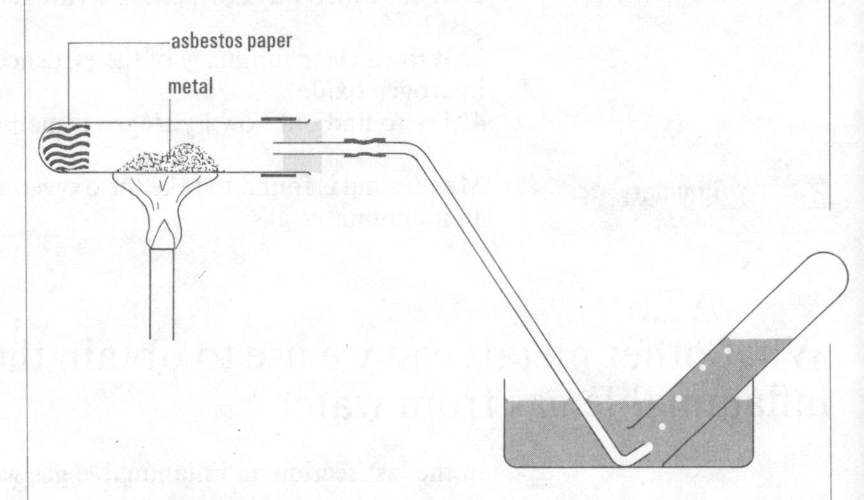

Loosely roll a strip of asbestos paper about 3 cm wide and put it at the bottom of the hard-glass test-tube. Using a teat pipette add as much water as the asbestos will absorb (about 2 cm^3). Clamp the test-tube horizontally and put one measure of iron powder midway along it. Assemble the rest of the apparatus and put a test-tube full of water over the end of the delivery tube.

Heat the iron powder gently at first and then more strongly, using the Bunsen flame also to warm the wet asbestos and generate steam.

Once you have started heating do not stop – why not?

Collect at least one and a half test-tubes of gas. Before you stop heating remove the end of the delivery tube from the water – why?

Try bringing the mouth of a test-tube full of gas (mouth downwards) to a Bunsen flame and then a tube half-filled with gas and half with air (use a half-filled tube and let the water drain out cautiously before putting in the cork) to a flame.
What happens:
To the pure gas?
To the mixture of gas and air?
What do you think would be the action of steam on heated zinc powder?
Why?
Try it and see if you are right.

In the experiment the pupils were asked to ignite a test-tube containing a mixture of the inflammable gas and air, when they will have heard a 'pop' or 'squeak'. Tell them that the gas which gives this 'pop' is hydrogen. If however the gas has been identified as hydrogen already (for example, after the magnesium–steam reaction) it can be called hydrogen all through this section.

Suggestions for homework

1. Tabulate the results of all your experiments with metals so that a direct comparison can be made of their order of reactivity towards oxygen and towards water.
2. A friend of yours tells you that he is going to build a steam engine using iron for the boiler tubes. Write him a letter explaining to him why this is not a good idea and suggesting alternative materials.

Summary

Pupils now know that hydrogen can be obtained from water, and that the order of reactivity of metals towards oxygen appears to be paralleled by an order of reactivity towards steam.

A7.4
A more convenient way of preparing hydrogen in the laboratory

Pupils are now shown a more convenient way of making hydrogen, using zinc and dilute sulphuric acid. A stream of hydrogen is then burned, in order to test the suggestion that water is hydrogen oxide.

A suggested approach

Objectives for pupils

1. Familiarity with a convenient apparatus for preparing gases in the laboratory
2. Knowledge of water as the oxide of hydrogen
3. Awareness of some of the history of hydrogen (Boyle and Cavendish)

The method used by the pupils in A7.3 was sufficient to provide small quantities of hydrogen, but will not provide the larger quantities that are sometimes needed. Tell the pupils that nearly 300 years ago Robert Boyle discovered another way of producing hydrogen. He found that some metals reacted with acids to produce hydrogen. Once again the pupils can use the reactivity series to predict which metals are most likely to do this. They may try small quantities of all the metals they have come across (*except sodium*) with dilute sulphuric acid in test-tubes to see how they react. Having done this the teacher should prepare hydrogen on a larger scale as described below. This method was used by Henry Cavendish nearly 100 years later when he made a thorough investigation of the properties of the gas.

Experiment A7.4a

Apparatus

The teacher will need:

Filter flask, fitted with delivery tube, bung, and tap funnel

Taper for igniting the hydrogen

(Continued)

Preparing hydrogen

***Caution.* Under no circumstances light the hydrogen as it comes out of the delivery tube.**

Procedure

Place a few pieces of granulated zinc in the filter flask, add a few drops of copper sulphate solution, and assemble the apparatus as in the diagram overleaf. Explain here that the copper sulphate 'speeds up the reaction'. Sulphuric acid (1 M) is poured onto the zinc *via*

Crystallizing dish or small trough

4 test-tubes, 150 × 25 mm, and corks

1M sulphuric acid

Zinc, granulated

Copper sulphate solution

Detergent solution

the tap funnel and the hydrogen collected in test-tubes over water in a crystallizing dish or small trough. The first two or three test-tubes full will of course contain displaced air.

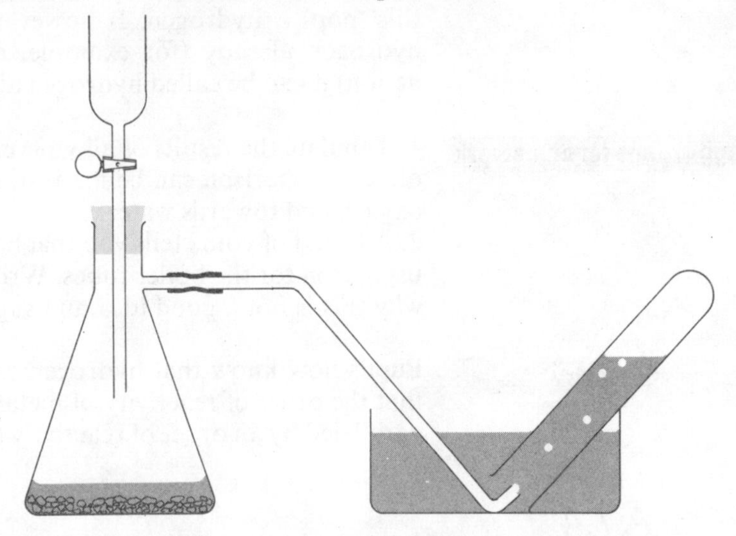

Having found that the gas can be ignited (thus revealing its presence) the teacher should do 'pouring' experiments to find out its density. On pouring upwards into dry test-tubes and igniting, the hydrogen will be mixed with air and will explode on ignition with the characteristic 'pop'! A 'mist' on the inside of the test-tube will again be observed.

Soap bubbles may be blown by inverting the delivery tube and dipping the end into detergent solution. The bubbles will rise rapidly, again demonstrating the low density of the gas. Tell the pupils that it is the least dense gas known.

If water is really hydrogen oxide, should we not be able to make it by burning hydrogen? In the next experiment this is put to the test. Note that dangerous explosions can take place if mixtures of hydrogen and air are ignited; this experiment **must** be done by the teacher, and a safety screen must be placed between the apparatus and the pupils. Hydrogen should be taken from a cylinder of the gas, and **not** from the apparatus of the previous experiment, which may contain some air.

Experiment A7.4b

Is water formed when hydrogen is burned?
This experiment **must** be performed by the teacher.

Apparatus
The teacher will need:

Plastic safety screen

U-tube or calcium chloride tube

Thistle funnel with stem bent as in diagram, fitted with bung for the test-tube

(Continued)

Procedure
The hydrogen is first dried by passing it through a tube containing silica gel or anhydrous calcium chloride. When all air has been displaced completely from the apparatus the hydrogen is burnt at a metal or ceramic jet. The hydrogen flame should be about 3 cm high and be allowed to burn beneath the thistle funnel as indicated in the diagram. The gas formed by the combustion is drawn through the apparatus by a filter pump, and a liquid soon condenses in the cooled test-tube. After about ten minutes 2 or 3 cm^3 of liquid will

Test-tube with side arm, 125 × 16 mm

Beaker, 250 cm^3

2 stands and clamps

Filter pump and connection tubing

Metal or ceramic jet and connection tubing

Test-tube, 100 × 16 mm

Thermometer, −10 to +110 °C

Bunsen burner and asbestos square

Hydrogen cylinder

Silica gel or anhydrous calcium chloride

Anhydrous copper sulphate

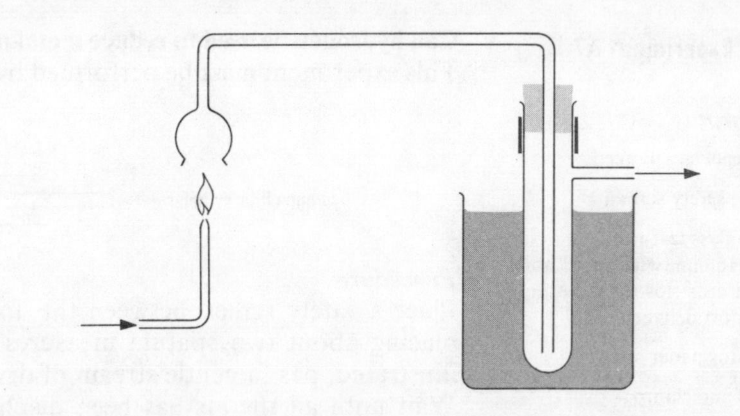

have collected. The nature of the liquid should be investigated by a boiling point check and the reaction with anhydrous copper sulphate.

Suggestion for homework

Chemists use analysis (breaking down) and synthesis (putting together) to investigate substances. Explain how you have used these processes to investigate water.

Summary

Pupils in this section have seen the laboratory preparation of hydrogen, know it to be a gas of low density which explodes when mixed with air and ignited, and have seen that water is formed when it is allowed to burn in a controlled manner.

A7.5
Can hydrogen be used to reduce metal oxides?

A suggested approach

The position of hydrogen in the reactivity series is used to predict that it can reduce lead and copper oxides. The prediction is confirmed by experiment.

Objectives for pupils

1. Ability to use an order of reactivity to predict reactions
2. Knowledge of the position of hydrogen in the reactivity series
3. Knowledge of some properties of hydrogen, including its uses in industry
4. Recognition of the usefulness of electricity to test substances

Ask the pupils to consult the reactivity series and see which metal oxides might be reduced by hydrogen. If the metals themselves were used to remove oxygen from water then it is unlikely that hydrogen will be able to remove oxygen from these metal oxides. (The case of iron in which the reaction can be reversed by altering the conditions should be mentioned now to avoid confusion later.) Thus from the reactivity table it seems likely that mercury, lead, copper, and iron oxides could be reduced by hydrogen, whereas the oxides of sodium, calcium, magnesium, and zinc could not. The experiment may be tried with copper and lead oxides, and one oxide, such as zinc oxide, which will not be reduced with hydrogen.

Experiment A7.5

Can hydrogen be used to reduce metal oxides?
This experiment **must** be performed by the teacher.

Apparatus

The teacher will need:

Plastic safety screen

Hard-glass test-tube, 125 × 16 mm, with small hole blown near closed end, bung and short delivery tube

Asbestos paper strip

Stand and clamp

Bunsen burner and asbestos square

Taper

Rubber tubing (to connect delivery tube to hydrogen cylinder)

6 V bulb and holder

6 V battery or alternative d.c. supply

Connecting wire

Hydrogen cylinder

Lead(II) oxide

Copper(II) oxide

Zinc oxide

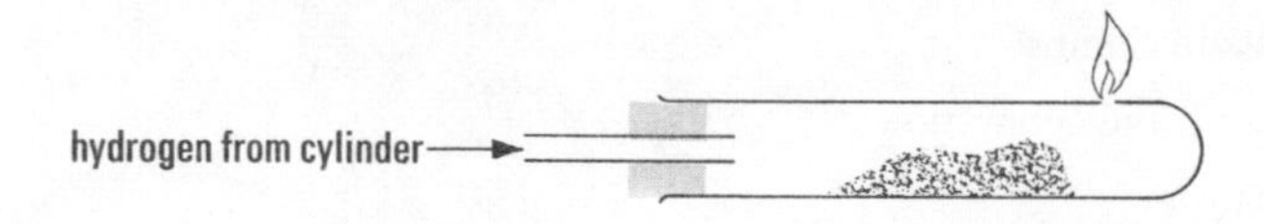

Procedure

Place a safety screen between the apparatus and the class. After placing about two spatula measures of oxide in the test-tube as illustrated, pass a gentle stream of dry hydrogen through the tube. Wait until all the air has been displaced, then light the escaping hydrogen at the hole and turn down the supply so that the flame is only about 2 cm high. The tube requires only gentle warming for the reduction to take place. Notice that in the case of copper oxide, a glow spreads through the powder as reduction occurs. Draw the pupils' attention to the condensation of moisture near the end of the tube.

To do this, do not place the oxide too near the hole end of the test-tube. In the case of the lead oxide, place the oxide on a piece of asbestos paper, previously heated in a Bunsen flame to dry it, so that the beads of lead formed do not attack the glass tube.

To confirm that the residual powder is indeed a metal, it can be shown to conduct electricity by including it in a circuit with a bulb and battery. Compare this conductivity with that of a sample of the oxide.

Suggestions for homework

1. Write word equations for the reactions of hydrogen with metal oxides.
2. What is hydrogen used for in industry? Make a list of as many uses as possible. Which uses do you think are the most important? (*Note.* This assumes that pupils have access to a library.)
3. Make a summary of the work you have done in this Topic.

Summary

Hydrogen is now seen to be a reducing agent for oxides of metals low in the reactivity series.

Topic A8

The effects of electricity on substances

Purposes of the Topic

The purpose of this Topic is to see if electrical energy can be used to decompose substances, that is, to bring about chemical changes, just as heat energy can. To do this the pupils are introduced to simple methods for testing substances for electrical conductivity, and they see that chemical changes take place at the electrodes when certain molten substances, and some solutions, conduct electricity. It is not intended that any theoretical explanation for the observations should be attempted at this stage.

The experience gained in this Topic will be the basis of developing important concepts later in the course but at this stage the emphasis is on the general aims that students should (1), 'Acquire basic knowledge about the behaviour of substances', and (4), 'Develop manipulative skill in laboratory procedures with common apparatus'.

Contents

Timing

This is an important Topic and plenty of time should be allowed in the laboratory for the necessary experiments to be done. A8.1 will need two or three double periods and A8.2 will need one or two. Thus a maximum of five double periods should leave plenty of time for the discussion of results. If time is short it should be possible to cut down the number of substances tried in A8.1 and in A8.2 and complete the Topic in two double periods.

Introduction to the Topic

In this Topic electrolysis is explored first in molten electrolytes and then in solutions. The order is deliberate so that the pupils will see that in adding water a change of behaviour can occur. Thus water plays an *active* and not a passive role in the electrolysis of solutions.

By the end of this Topic the pupils should know how to set up circuits for electrolysis. They should realize that metals can be expected to appear at the negatively charged electrode, and non-metals at the positively charged electrodes. They should be able to use the words anode, cathode, electrode, and electrolyte with confidence and should be able to handle a battery (or other L.T. supply) with the same skill with which they have learnt to use a Bunsen burner.

No theoretical explanation of electrolysis is given until Stage II.

Alternative approach

Another treatment of this subject can be found in Alternative B, where the use of electricity in getting new materials is met earlier (Topic B6) than in this scheme. Here, the idea of elements comes before electrolysis, and can be used to help the understanding of the latter, to the extent that electricity can be seen to split up compounds. In Alternative B electrolysis precedes the idea of elements and is used to help to define them. Topic B6 also includes an elementary consideration of simple cells, not mentioned in this Topic.

Background knowledge

The use of electricity in separating elements from compounds was first met in Topic A1, section A1.6, where copper was obtained from copper sulphate by electrolysis. Electrical conductivity was used in Topic A5 as one property which helped to classify elements into metals and non-metals.

Subsequent development

The subject is taken further in Stage II.

Further references

for the teacher

Additional experiments suitable for this Topic are given in *Collected experiments*, Chapter 5 'The effect of electricity on substances'. A general discussion on the treatment of electrochemistry in Stage I is given in the *Handbook for teachers*, Chapter 6; there are also some suggestions on the interpretation of experimental observations in Chapter 7, pages 81–82.

Reading material

for the pupil

Study sheet:
Chemistry and electricity helps pupils realize that the batteries they use in torches and transistor radios originate from the observations and experiments made by two Italians over 170 years ago. The Study sheet then deals with the uses of electrolysis and with other ways to make electric current, which have led to our enormous consumption of electricity today.

A8.1
Which substances conduct electricity?

In this section the pupils use a simple apparatus to find out whether the substances they have been given conduct electricity (1) in the solid, and (2) in the molten state. They note any changes which take place in the substance when a current passes.

A suggested approach

Objectives for pupils

1. Familiarity with an electrolysis technique for molten substances
2. Knowledge of the behaviour of substances in electrolysis
3. Understanding of the meaning of the words: electrode, anode, cathode, electrolyte

It is probably wise to start the first lesson on this Topic by reminding the class of the last experiment in which they used electrolysis. This was the experiment in which copper was extracted from copper sulphate solution (A1.6). They saw then that a battery is a source of electrical energy much as a Bunsen burner is a source of thermal energy. In either form, energy can help to bring about chemical changes.

To find out more about the effect of electricity on substances, suggest to the class that they need to make a systematic investigation. Show them a selection of substances (see the list below) and say that we are to find out whether these substances conduct electricity (1) as they are, and (2) when they are molten. Ask them to note any changes which take place when the electricity is passing through. They will have to distinguish carefully between this effect and the effect of the heat when the substances are being melted. A range of substances to be heated is given below. Introduce the words electrode, anode, cathode, and electrolyte.

If each pupil is to try all the substances suggested a considerable degree of organization is necessary. It is probably better to adopt a compromise arrangement whereby the work is divided between the pupils and the teacher. In this case the teacher could test all the molten materials, leaving the pupils to test them at room temperature. In any case it is wise for the teacher to demonstrate the tests on lead iodide and potassium iodide, as these compounds have high melting points. One good way of organizing the class is to set out a number of 'stations' at which perhaps two or three substances for testing are to be found, and have the pupils go from one to another, until they have tested all the samples available. This method has the advantage of requiring much smaller amounts of materials. A d.c. supply and appropriate electrodes should be at each station.

Experiment A8.1

Finding out which substances conduct electricity

Apparatus

Each pupil (or pair) or the teacher will need:

Experiment sheet 18

6 V bulb and bulb holder

6 V battery or alternative d.c. supply

Carbon or steel electrodes and holder

Connecting wire, 2 pieces fitted with crocodile clips

Bunsen burner and asbestos square

Tripod and pipe-clay triangle or stand and clamp

Small pieces of lead, copper, and other metals

Crucibles or hard-glass test-tubes, 150 × 25 mm, containing samples of naphthalene, sulphur, polythene, wax, sugar, lead(II) iodide, and potassium iodide

Procedure

Each experiment should have a circuit to contain the electrodes, bulb, and holder using the low voltage supply available. The bulb is included to indicate when current is flowing.

Each of the samples in turn should be tested to see if it conducts electricity by holding the electrodes in contact with it.

If time does not permit all the substances to be tested by every pupil, results from all the class can be pooled at the end of this part of the experimental work when the class is called together for discussion. Only the metals (and the carbon electrodes) appear to conduct electricity.

Samples of the non-conductors should be heated until they just melt, to find out whether the molten material conducts electricity. In this case the electrodes are more conveniently supported in the electrode holder while the investigation is carried out. The samples should not be overheated because some substances will burn if heated too much (for example, sulphur, wax, naphthalene, and polythene). Electrodes must be scraped clean before testing the next substance, unless the 'station' approach described above is used, in which case the electrode stays with the same sample all the time. In addition to those substances which conduct electricity when solid, it will be found that potassium iodide and lead(II) iodide conduct when molten.

Experiment sheet 18

You have studied already the effects of heat on a number of substances, and investigated the reactions which occurred. You are now going to look at the effects of passing an electric current through a number of substances.

Experiment sheet 18 *(continued)*

You are going to test various chemicals to see whether or not they conduct electricity. You will already have discussed the pieces of equipment you are going to use and these are not shown in detail. The diagram is only a scheme to show how the different items are to be connected.

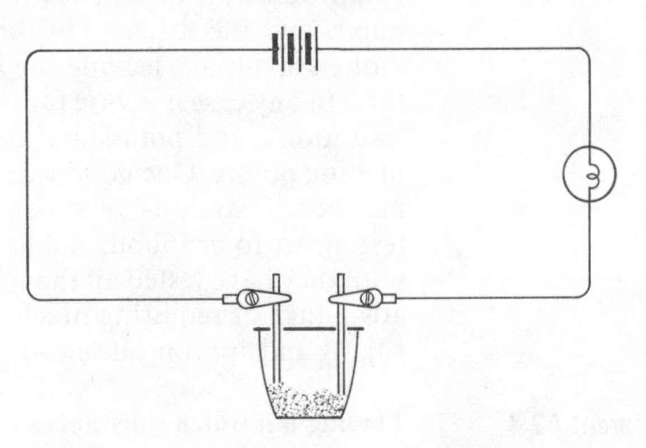

The way to find out whether the solid substances conduct electricity is to hold the two electrodes firmly against each substance in turn and see if the lamp lights up.

Enter the names of the substances in the table below as you test them, and put a tick in the appropriate column.

Now try to find out what happens when the substances are melted and the electrodes inserted in the liquid. Put two or three measures in a crucible supported on a pipeclay triangle on a tripod and heat them with a medium Bunsen flame.

In the table below put an L against those substances that conducted when molten although they did not conduct when solid.

Substance	Conductor	Non-conductor

New words

Electrode
Anode
Cathode
Electrolyte

Suggestion for homework

Read the Study sheet *Chemistry and electricity*.

Summary

Pupils have now seen how to test both solid and molten substances for electrical conductivity. They will realize that amongst the solids only metals and carbon (graphite) conduct electricity, but that certain solid non-conductors conduct electricity when they are melted. They should know that these substances are called electrolytes, and will probably have noticed that chemical changes take place at the electrodes in these cases.

A8.2
Do solutions conduct electricity?

The pupils now investigate the effect of an electric current on a number of solutions. They find that only some of these conduct electricity, and they learn that water can play a part in electrolysis.

A suggested approach

Objectives for pupils

1. Familiarity with an electrolysis technique for aqueous solutions
2. Knowledge of the behaviour of certain aqueous solutions on electrolysis, including the part played by the water

The last experiment showed the pupils that there is a group of substances, electrolytes, which conduct electricity when molten and which decompose as they do so. 'Do substances conduct electricity in solution?' Discuss with the class how an answer to this question can be found and devise an experiment to find out (1) whether the solutions chosen do conduct electricity, and (2) what happens if they do so.

Experiment A8.2

To find out whether water or solutions conduct electricity

Apparatus

Each pupil (or pair) will need:

Experiment sheet 19

6 V bulb and bulb holder

6 V battery or alternative d.c. supply

Carbon electrodes and holder

Connecting wire, 2 pieces fitted with crocodile clips

Beaker, 100 cm^3

Paper tissues or clean cloth

Distilled water

Access to 100 cm^3 beakers containing approximately M or 0.5M solutions of:

Potassium iodide

Sulphuric acid

Sodium hydroxide

Sodium chloride

Copper sulphate

Zinc sulphate

Cane sugar

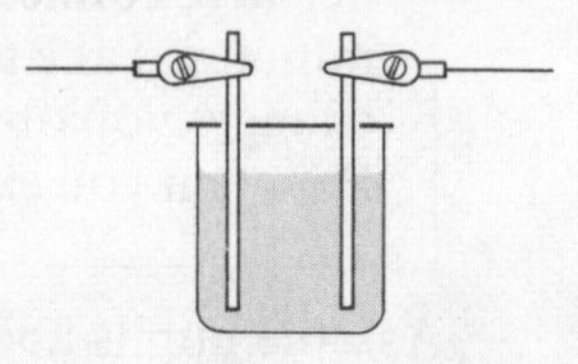

Procedure

The pupils should arrange a circuit to contain the electrodes, bulb, and holder, using the low voltage supply available. The bulb is included to indicate when current is flowing.

An experiment to find out if distilled water appears to conduct electricity should be done first since this has been used to prepare the other solutions.

They should dip the electrodes into a solution selected from those above to see if it conducts electricity, and note the result. After cleaning the electrodes they should be wiped dry, and the procedure repeated with other solutions. It will be found that, except for the sugar solution, they all conduct electricity.

As with Experiment A8.1 it may be found more convenient to have a number of stations at which one or two solutions can be tested, and let the pupils move from one to the other, rather than giving each pupil samples of all the solutions.

Experiment sheet 19

Having studied the effect of an electric current on some solids and melted substances you can now deal with water and solutions in water.

The circuit is the same as that used in the previous experiment (*Experiment* 18). A small beaker is used to hold the

Experiment sheet 19 *(continued)*

liquids being tested. Try distilled water first, then tap water and the other solutions provided. Make sure that a clean beaker is used each time and that the electrodes are washed before a new solution is used. Make a table below to record your observations; it will be helpful to note the relative brightness of the bulb in the table.

When a solution conducts electricity, something else always happens. Can you see what that is?

It will be interesting to find out more about what is happening in these experiments. Let us have a closer look at what happens when the copper sulphate solution conducts electricity.

Use carbon electrodes and leave the current on for a definite time – say two minutes. Take the electrodes out. What changes do you see on:
1. The positive electrode?
2. The negative electrode?

Repeat the experiment, but change over the positive and negative connections. Again leave the current on for two minutes. Take the electrodes out and again make a note of any changes you observe.
How can you explain these?

Ask the pupils about the significance of the experiments they have performed. Some of the solutions conduct electricity and some do not. Tell them that substances which conduct electricity in solution are also called electrolytes. 'Has water played a part?' Obviously it has played some part as these substances do not conduct electricity in the solid state. One of the substances chosen, potassium iodide, was electrolysed in the molten state and in solution. What difference did the pupils notice in its behaviour? They will have seen that in the case of potassium iodide in solution a gas is given off at the cathode. Some of them may even have identified it as hydrogen. In any case the result is different when water is present and therefore water must play a part.

Suggestions for homework

1. Write a summary of what you have learnt in this Topic.
2. Explain why there is reason to believe that water may play a part when solutions of electrolytes are electrolysed.

Summary

Pupils have now seen how to test aqueous solutions for electrical conductivity and have noticed that chemical changes take place at the electrodes. They will have found that zinc and copper form at the cathode during the electrolysis of zinc and copper sulphates, but bubbles of gas are seen at the cathode when potassium iodide and sodium chloride are tested, and not the metals. This suggests that the water plays some part in the electrolysis.

Topic **A9**

Chemicals from the rocks

Purposes of the Topic

The main purpose of this Topic is to investigate certain minerals to try to discover their composition, and to see how some valuable elements can be obtained from them. This work will provide the pupils with opportunities to plan and carry out investigations; at the same time they will be made aware of the usefulness of chemistry in the industrial world, and will obtain some knowledge of the industrial chemistry of iron, copper, and limestone. These activities are in line with the general aims (page 13) that pupils should (4) 'Develop manipulative skill in laboratory procedure with common apparatus', (5) 'Gain confidence in learning by discovery', and (7) 'Find out about the sources and uses of the substances they meet, and any hazards connected with them'.

Contents

A9.1 How can iron be extracted from iron ore?
A9.2 Investigation of a mineral: malachite
A9.3 Investigation of limestone

Timing

The first section will probably need only one double period, whereas the other two will certainly need two each. Allowing one more double period for discussion, a maximum of six double periods should enable the Topic to be studied thoroughly. There is no very satisfactory way of cutting this time down, but if time is very short it is probably wise to concentrate on A9.3 and omit A9.1 and A9.2. Alternatively concentrate on A9.2 which is an interesting investigational exercise and omit A9.1 and A9.3.

Introduction to the Topic

After nearly two years of chemistry the pupils have learnt enough about the way in which chemical apparatus can be handled, and the way in which some simple chemicals behave, to embark on more adventurous investigations. In this Topic, therefore, the pupil is encouraged to find out more for himself and to suggest his own methods and experiments. Running parallel to this development is the theme of the usefulness of chemistry. All the chemical changes investigated in this Topic are important in the world outside and this is one of the points in the course where practical applications of chemistry are particularly stressed. By the end of this Topic the pupils should be aware of this usefulness in the field of iron and steel manufacture, copper, and of the chemistry of chalk and its derivatives.

Supplementary materials

Film loops
1–8 'Iron extraction'
1–10 'Limestone'

Films
'Limestone in nature'

'Study in steel'
'The big mill'
BBC programme 'Corrosion'
See Appendix 4 for brief descriptions and further details of these films

Study sheet:
Chemicals and rocks aims to give pupils an idea of the great variety of rocks and minerals, and to show how chemistry can be used to study them.

Reading material

for the pupil

Useful books include:
Schools Council Integrated Science Project (1973) *Rocks and minerals.* Longman/ Penguin Books.
Brennand, S. (1973) *Gold and granite.* Puffin Books.
Allan, M. (1972) *What do we know about the Earth.* Blackie.

A9.1
How can iron be extracted from iron ore?

This section brings out the practical usefulness of chemistry through an investigation into the problems of isolating iron from iron ore. The reactivity series is used to suggest the idea of carbon as a reducing agent.

A suggested approach

Objectives for pupils

1. Awareness of the usefulness of chemistry
2. Familiarity with planning an experiment
3. Knowledge of the industrial extraction of iron

The teacher will need

Film loop projector

Film loop 1–8 'Iron extraction' (with notes)

Sample of iron ore

Useful films

'Study in steel'

'The big mill'

BBC programme 'Corrosion'

See Appendix 4 for brief descriptions and further details

The first lesson in this Topic may start with a discussion on the usefulness of chemistry. One of the largest areas of practical chemistry has always been the extraction of materials from mineral sources – 'Chemicals from the rocks'. If there are mines or mineral deposits of any sort in the locality of the school they may be used as examples here. In any case remind the class that they found that copper could be extracted from copper pyrites when they studied Topic A1. The question to be answered by experiment now is: 'Can the iron be extracted from iron ore in a similar way?' Draw the attention of the class to the reactivity series and ask them to suggest a means of reducing iron oxide. If they suggest magnesium the reaction may be demonstrated, but *take care* – the iron oxide must be absolutely dry. Apart from the violence of the reaction, magnesium would be an expensive element to use when extracting iron from an iron oxide. Ask the class for further suggestions. Carbon is an obvious answer – the pupils know its position in the reactivity series from their experiments in section A6.2. How can this be tried out? A simple technique is described below.

Experiment A9.1

Trying to get iron from iron ore

Apparatus

Each pupil (or pair) will need:

Experiment sheet 20

Procedure
The pupils mix together a spatula measure of iron(III) oxide with a rather greater amount of powdered wood charcoal. They heat the mixture strongly in a hard-glass test-tube and allow it to cool. A

The teacher is discussing the film loop 'Iron extraction' during the second showing.

black magnetic product is obtained which evolves hydrogen on treatment with 1 M sulphuric acid. (*Experiment sheet* 20 gives fuller details and is reproduced below.)

Note Some samples of iron(III) oxide are magnetic and some become magnetic on heating. The magnet test alone is therefore not reliable evidence that iron is present.

Bunsen burner and asbestos square

Hard-glass test-tube, 75 × 12 mm

Test-tube holder

Magnet

Spatula

Iron(III) oxide

Powdered wood charcoal

M sulphuric acid

Experiment sheet 20

You will remember that you have already seen how certain metals are obtained from their ores. Look back in your book to remind yourself of the experiments you did. Write down the names of two metals you have studied yourself and the names of their ores.

How were the metals obtained from the ore in each case?

We will try to obtain iron in the laboratory by the reduction of iron oxide. Look at the place of iron in the affinity list in your notebook; will it be easy to reduce iron oxide to iron? What reducing agent might do it?

Grind up a measure of iron ore (or iron oxide from the bottle) with about two measures of powdered charcoal, in your mortar. Place some of the mixture on a piece of asbestos paper about 2 cm × 5 cm, cover it with a little more charcoal, and heat it very strongly in a roaring Bunsen flame. After five minutes, allow it to cool.

How can you tell whether you have succeeded in producing any metallic iron?

There are at least two tests you should be able to think of, and to try. What happens when you try them?

After the experiment discuss the results with the class. The experiment shows that iron oxide can be reduced by carbon. This result may then be used to introduce the subject of the industrial production of iron and steel from iron ores. Wall charts may be used to explain the way in which the blast furnace works and the film loop 1–8 'Iron extraction' should also be shown at this point.

Suggestions for homework

1. How many things made of iron or steel can you think of? Why do you think iron or steel is used rather than some other material such as wood?
2. What metals, other than iron and steel, can you find in your home? Make a list of the metals and explain why you think they are used for each particular purpose.
3. A friend of yours tells you that he is going to have metal window frames fitted in his house in place of wooden ones. Which metal would you advise him to use for this purpose and why?

Summary

Pupils have now considered the importance of the extraction of materials from minerals, and in particular, the extraction of iron from iron ore.

They should understand the principles of reduction of iron ore with carbon, and know something of the details of the industrial processes for the production of iron and steel.

A9.2
Investigation of a mineral: malachite

In this section the pupils take malachite as an example of a 'rock' or ore and try to find out what it is composed of. They now have enough chemical knowledge to find that it contains copper carbonate.

A suggested approach

Objectives for pupils

1. Familiarity with planning an investigation
2. Knowledge of the uses and simple chemistry of copper

The teacher will need

A piece of malachite

At this stage in the course the pupils have enough experience to be able to investigate a problem of this kind with confidence. They will still need some guidance, however, and it is a matter of the teacher's skill to decide to what extent his help is needed for a particular class and for a particular pupil or pair of pupils. In planning this type of lesson the teacher has also to decide how far to go in discussion with the class as a whole and how far to leave it to individual discussion with the pupils once the experiment has started.

Show them a sample of malachite and ask them how they could try to find out what it contains. As malachite is expensive, it is wise to have one sample to show and thereafter to give the pupils 'ground up malachite', that is, copper carbonate powder, to experiment with. 'What is this substance and what does it contain?'

Discuss the problem with the class. How should they start? Two main lines of investigation may be suggested.

1. Some may suggest heating it. Let them try. It becomes black. What is that black substance? There are probably only two black substances they have come across; carbon and copper oxide. What should they do to find out which it is?

Let them try burning it in oxygen and testing for carbon dioxide; reducing it with a reducing gas or adding warm acid to it in a beaker and electrolysing.

2. Some may suggest adding acid to it. Let them try. What is the gas evolved? Let them test it. What is the blue solution? Some may suggest electrolysing it or dipping a clean nail in it.

After these two lines of investigation have been pursued by at least some of the class there will be enough experimental evidence to deduce that malachite contains copper carbonate. The experiments above are now described or referred to.

a. Heating malachite. The experiment should start as in A2.1. If the pupils suggest that something has been lost they should be encouraged to test by weighing (A2.3) and by testing for a gas. They will discover that carbon dioxide and water are both driven off during heating.

b. Burning the black powder. This can be done in the manner of Experiment A4.1. A sample of the black powder should be heated in a combustion spoon and then lowered into a test-tube of oxygen.

c. Reducing the black powder. This can be done (by the teacher) as described in Experiment A7.5.

d. The action of acid on malachite. This can be done in a test-tube using one spatula measure of malachite and one teat pipette full of 2 M hydrochloric acid. The gas can be tested by withdrawing portions with a teat pipette and expelling it through limewater.

e. Electrolysing the blue solution. See A1.6b.

Experiment A9.2

An investigation of malachite

Apparatus

Each pupil (or pair) may need:

Hard-glass test-tube, 125 × 16 mm, with a small hole near the closed end, or a combustion tube

Bung or cork carrying a short length of glass tubing

Rubber tubing to connect town gas supply to glass tubing

(Continued)

There is no *Experiment sheet* for this experiment, as the course that the experiment takes will be dictated largely by the suggestions of the pupils.

If experiments of the type of those described as (*a*) to (*e*) above are done, experiment (*a*) will show that carbon dioxide can be obtained from malachite by heating it, and experiment (*c*) that copper can be obtained from it. The carbon dioxide is also given off when malachite is treated with acid, experiment (*d*). In discussion with the class we may then deduce that malachite contains copper, carbon, and oxygen. They will not know at this stage that the action of acids

Stand and clamp
Bunsen burner and asbestos square
Taper or splint
6 V bulb and bulb holder
6 V battery or alternative d.c. supply
Combustion spoon and oxygen supply
Hard-glass test-tube, 125 × 16 mm
Crystallizing dish
Teat pipette
Spatula
2 test-tubes 100 × 16 mm
Copper carbonate
M sulphuric acid
Limewater
Iron nail

on carbonates (and hydrogen carbonates) gives carbon dioxide; they must be told this and then led to the conclusion that malachite is a copper carbonate.

The way in which this lesson is handled will depend very much on the teacher himself and on the way in which the pupils react. Thus one pair may think that they have carbon and another may see quickly that it is copper oxide. In the same way a slow class may need a great deal more prodding than a more intelligent one. The success of the lesson depends on the ability of the teacher to make his approach flexible.

Suggestions for homework

1. Read the Study sheet *Chemicals and rocks*.
2. Find out all you can about copper and its uses.
3. Find out which elements are obtained from their ores commercially by electrolysis. Whereabouts are these elements to be found in the reactivity series?

Summary

Pupils have had an opportunity to plan an analytical investigation, and should have recalled a number of chemical reactions met in previous Topics. After this section of work they should know something of the industrial production of copper and its uses.

A9.3 Investigation of limestone

Malachite was investigated in the last section; in this section the rock to be investigated is limestone (or chalk or marble). In this case the fact that the products are all white makes the investigation more difficult and an analogy with malachite is helpful.

A suggested approach

Objectives for pupils

1. Familiarity with planning an investigation
2. Knowledge of the uses and simple chemistry of limestone

The teacher will need

Film loop projector

The problem which confronts the class now is very similar to that of the last section and the lesson might therefore start with a reminder of the procedure adopted in investigating malachite. The problem now is to analyse another substance. This may be chalk, limestone, or marble. Indeed one way to approach this lesson is to give the pupils different substances to analyse (some have chalk, some marble, and some limestone) and let them come to the conclusion that they are all forms of the same compound. If the school is in a district where any of these three substances are found, the pupils may be asked to bring in samples for analysis.

Film loop 1–10 'Limestone' (with notes)

Useful film

'Limestone in nature'

See Appendix 4 for a brief description and further details

In the initial discussion with the class it is very likely that the pupils will suggest either heating the substance or treating it with acid as in the case of malachite. Allow them to try this and ask them what they would expect to see if a chemical change took place in either of these cases. A series of tests is included in the experiment described below.

Experiment A9.3

An investigation of another mineral

Apparatus

Each pupil (or pair) will need:

Experiment sheet 21

Loop of fairly stout iron wire

Teat pipette

Beaker, 100 cm^3, and watch-glass

3 hard-glass test-tubes, 100 × 16 mm

Filtration apparatus or centrifuge

Nichrome wire

Bunsen burner and asbestos square

Stand and clamp

A few marble chips, pieces of limestone, and chalk

Concentrated hydrochloric acid

2M hydrochloric acid

A few calcium turnings

Limewater

Procedure

The suggestions made by the pupils are likely to include the following tests; in discussion the teacher can rearrange the order of the suggestions and fill in any gaps by suitable questioning until the following sequence of experiments is established. Results that the pupils will obtain are included in brackets after each suggested test.

1. Is the powdered mineral soluble in water? (No.)
2. What happens if it is heated? (No obvious change on heating in a test-tube, but some loss in weight found if a lump of the mineral is held in a piece of iron wire and heated directly by a roaring Bunsen burner flame.)
3. What happens if acid is added? (A gas is evolved which turns limewater milky; knowledge gained in Experiment A9.2 suggests that the mineral must be a carbonate. Note that the reaction with dilute sulphuric acid, used in Experiment A9.2, is very slow; dilute hydrochloric acid should be used here.)

In *Experiment sheet* 21, reproduced on the next page, the pupils are then led to an investigation of the heated mineral, to establish the relationship between calcium carbonate, calcium oxide, and calcium hydroxide. This goes along the following lines.

They examine the heated substance to see if it has changed. It looks much the same. They then see if it is still insoluble in water. If a little water is added they may notice that it gets warm and begins to crumble. This did not happen to the original substance. It must have been changed by the heating. By analogy with the malachite, the residue after heating might be an oxide, since the carbon dioxide has been driven off. Adding water to it makes a new compound called a 'hydroxide'. They next dry a sample over a steam-bath and then heat it more strongly. Water is evolved, suggesting that the slaked lime is a compound.

Water is again added to the residue and this time a solution is made and filtered or centrifuged to get some clear 'limewater'. To test the idea that this has been formed from the limestone by loss of carbon dioxide, they are asked to pass carbon dioxide through the solution. An insoluble white substance is formed.

By this time it will be clear that limestone, marble, and chalk are all behaving in the same way. It remains to find out what metal is present in these compounds.

In *Experiment sheet* 21, pupils are told that chalk, limestone, and marble are forms of a compound called calcite. It is then suggested that calcium may be the metal present in the material under investigation.

Now give the pupils some calcium metal and let them try to make the compounds from it, that is, burn a small piece of calcium again, add water, filter, and pass carbon dioxide through the filtrate. They then have some evidence for linking calcium with limestone, chalk, and marble.

Experiment sheet 21

You will be given some chalk or limestone or marble to investigate. The results of the previous experiment with malachite will help you in this investigation.

1. The mineral you have been given is. . . .
Examine it and describe its appearance.
Is it hard or soft?
Does it dissolve in water?
Heat a piece of it in a hard-glass test-tube. Do you notice any change?

2. Weigh a small piece of the mineral, about 1 cm across. It weighs . . . g. Support it in a loop of stout steel wire, held in a clamp on a stand, and heat it as strongly as you can in a Bunsen burner flame for ten to fifteen minutes. Leave it to cool.

3. While waiting, place another piece of the mineral in a test-tube and add about 2 cm^3 dilute hydrochloric acid to it.
Is a gas given off?
If so, try to find out what the gas is. Describe how you did this.
What do you think the gas is?
What does this suggest about the chemical nature of the mineral?
What sort of chemical compound might it be?

4. The heated mineral should be cool by now. Check this.
Weigh it again. It weighs . . . g.
Was there a gain or loss in weight?
The substance you have obtained is called *quicklime*.

5. Put the piece of heated mineral in a small dish (or a hard-glass watch-glass) and slowly add four drops of water from a teat pipette. Feel the outside of the dish cautiously with the palm of your hand. What do you notice?
What does this suggest to you about the action of water on the heated mineral?
Add six more drops of water. The quicklime has now been changed to *slaked lime*.

6. Dry your sample of slaked lime by placing it, in the dish or watch-glass, over a beaker of boiling water. When it is quite

dry to the touch, transfer some of it, about half, to a dry test-tube and heat it. What do you notice on the upper cool part of the test-tube?
What does this tell you about slaked lime?

7. Shake a little of the dry slaked lime with half a test-tube of distilled water in order to try to dissolve it. Filter (or centrifuge) to get a clear liquid. Pass a slow stream of carbon dioxide through this solution (either from a generator of the gas or by blowing through a glass tube). What happens?
It looks as though this is a solution you have met before. What is it?

8. You now have a problem; before you can write down the chemical names for chalk (or limestone or marble), quicklime, and slaked lime you need to know what else these substances contain. Malachite was found to be a compound of a metal whose name you know. The white substance that you have been studying may also contain a metal. It is not easy to extract the metal from it but if you are told that it is a form of a compound called calcite, you might guess that the metal is calcium. If you start from some calcium metal, you might solve your problem from the other end, by making compounds from the metal.

Support a calcium 'turning' in a loop of stout steel wire in a stand, with a tile or asbestos mat underneath to protect the bench, and heat it very strongly until it burns. What would you expect to be formed when calcium burns in air?

When the product is cool, investigate the effect of adding a few drops of water to it, as you did for the quicklime in (5) above. Add the product to more water in a test-tube, shake the mixture, and filter (or centrifuge) to obtain a clear solution. Pass carbon dioxide through the solution; what happens?

You should now feel fairly confident in stating that:
The chemical name for chalk and limestone and marble is . . . .
The chemical name for quicklime is . . . .
The chemical name for slaked lime is . . . .

After the experiment discuss the results and show film loop 1–10 'Limestone'.

Suggestions for homework

1. Chalk, limestone, and marble are all useful substances. Write an account of the uses to which they are put.
2. Make a short summary of the work you have done in this Topic.

Summary

Pupils have now had a further opportunity to plan an investigation. They should have a knowledge of the names of the various materials that they have met in the section – calcium carbonate, calcium oxide, and calcium hydroxide, the relationships between them, and the uses to which they are put. They are not expected to know the chemical formulae of the substances or the equations involved.

Topic **A10**

Chemicals from the sea

Purposes of the Topic

The purposes of this Topic are to investigate the materials present in sea water, and to learn something of the halogens.

Contents

A10.1 What chemicals can be obtained from the sea?
A10.2 What is the effect of electrolysing sea water?
A10.3 Getting iodine from seaweed
A10.4 A family of elements – the halogens
A10.5 The industrial production of the halogens

Introduction to the Topic

One double period will be needed for the first section and one each will probably be sufficient for the remaining sections. Five double periods should allow enough time for all the necessary introduction and discussion in the Topic.

In this last Topic in Stage I Alternative A, the pupils are introduced, for the first time, to a family of elements, the halogens. The elements appear during the course of an investigation into the nature of sea water. Chlorine and bromine are evolved at the anode during electrolysis experiments and iodine is extracted from seaweed.

By the end of this Topic the pupils should be familiar with chlorine, bromine, and iodine as a family of elements whose members have similar properties, and they should know something about their industrial production. They will have seen that calcium carbonate, calcium sulphate, sodium and potassium chlorides and bromides, magnesium chloride and sulphate may all be obtained from sea water. They will have learnt how to test for chlorides and sulphates as well as for magnesium, potassium, and calcium salts.

Subsequent development

The halogens are studied again in Stage II, Topic 12.

Further references

for the teacher

G. Fowles (1963) *Lecture experiments in chemistry* (6th edition, Bell), has a useful discussion on the composition of sea water and the way in which various chemicals can be obtained from it.

Supplementary materials

Film loops
1–11 'Fluorine manufacture'
1–12 'Fluorine compounds uses'
1–13 'Chlorine manufacture'
1–14 'Chlorine uses'
1–15 'Bromine manufacture'
1–16 'Bromine uses'
1–17 'Iodine manufacture'
1–18 'Iodine uses'

Film
'Chlorine'
See Appendix 4 for a brief description and further details

Reading material

for the pupil

Study sheet:
The halogens. Besides describing many of the beneficial ways we use these elements, the Study sheet questions the responsibility of the scientist who developed chlorine as a weapon in the First World War, and it points out the harmful effects of bromine in petrol. (The idea for the first part of this Study sheet is drawn from Holmyard, E. J. (1925) *An elementary chemistry.* Edward Arnold.)

A10.1
What chemicals can be obtained from the sea?

The evaporation of sea water reveals that it contains a number of important chemicals. The presence of chlorine and bromine in sea water is linked to that of iodine in seaweed, and this is used as a means of introducing the halogens.

A suggested approach

Objectives for pupils

1. Knowledge of some of the dissolved substances in sea water
2. Knowledge of tests for carbonates, chlorides, sulphates, calcium, magnesium, sodium, and potassium

The first lesson on this Topic may be opened with a discussion on the nature of sea water. The pupils will know that sea water tastes salty (salt water), but do they know what chemicals are actually found in the sea? A great deal of material produced by the weathering of the rocks is washed down into the sea, and we would therefore expect to find many substances in sea water. It is the object of this section of work to find out what substances we can obtain from sea water. One way to do this is to start with a litre of sea water and to evaporate it down in stages. There are several ways of handling this experiment. It may be done as a teacher experiment with the pupils assisting with the tests and looking at the crystals obtained under a microscope (or on a screen if a microprojector is available), but this can be rather dull for the pupils. A better way to go about it is to evaporate some water beforehand. One might then show the class a litre of sea water and explain that this is the quantity which has been evaporated down by various degrees. In each case some solids have appeared. Different pairs of pupils may then take sea water of differing concentrations, evaporate it down to the next stage and find out what substance is precipitated during the evaporation; or each stage may be reached beforehand.

The substances which come down must be identified. Some of the tests used will be familiar to the pupils from previous work. The limewater test for carbon dioxide was first met in section A4.3 and calcium was burnt in air in section A5.1 where its characteristic flame colour was noticed. The tests for chloride, sulphate, and magnesium (using the magneson reagent) should be carried out by the pupils before starting on the main part of the lesson.

Experiment A10.1a

The evaporation of sea water

This work should be carried out prior to the lesson. The time required is about two hours.

Apparatus

Beakers: 2 one-litre, one each 250 cm^3, 100 cm^3, 50 cm^3, and 25 cm^3 (graduated ones are helpful)

(Continued)

Procedure

Start with one dm^3 of sea water (500 cm^3 in each of the two one-litre beakers) and evaporate until the total volume is about 500 cm^3.

2 Bunsen burners and asbestos squares

2 tripod stands and gauzes

Buchner funnel, flask, and filter papers

Filter pump

4 crystallizing dishes

Filter tube and Hirsch funnel to fit

Transfer all the water to one of the beakers and evaporate to 125 cm^3. Filter under suction using the Buchner apparatus and transfer the small quantity of solid obtained to the first crystallizing dish.

Evaporate the filtrate to 25 cm^3 in the 250 cm^3 beaker. A great deal of solid is produced. When bumping begins at the 50 cm^3 mark, filter off as before and continue evaporating in a smaller beaker. Repeat when necessary. Transfer all the solid to the second crystallizing dish. Evaporate the filtrate to 10 cm^3 in the 50 cm^3 beaker and filter with suction using the filter tube and Hirsch funnel. Transfer the crystals to the third crystallizing dish.

Evaporate the filtrate to 5 cm^3 and filter as above. Transfer the solid to the fourth crystallizing dish and retain the small quantity of filtrate for use in Experiment A10.2b.

Once the samples of solids have been obtained as just described, the pupils can do a series of tests to identify them.

Experiment A10.1b

What dissolved solids are present in sea water?

Apparatus

Each pupil (or pair) will need:

Experiment sheet 22

8 test-tubes, 100 × 16 mm, in rack

5 cm length of nichrome wire in glass rod for flame tests

4 watch-glasses

Cobalt blue glass

Access to the following reagents:

Limewater

Silver nitrate solution acidified with 2M nitric acid

Barium chloride solution acidified with 2M hydrochloric acid

Magneson reagent

Concentrated hydrochloric acid

Distilled water

Solid sodium chloride

Solid potassium chloride (analytical reagent grade if possible)

Solid magnesium chloride

Solid calcium chloride

Solid sodium sulphate

(Continued)

Procedure

When the pupils have completed the investigations given in part (1) of *Experiment sheet* 22, they should be shown the chemical test for magnesium using the magneson reagent. This is carried out in the following manner. To 2 cm^3 of the test solution add one drop of 0.5% solution of magneson in water, followed by 2–3 cm^3 of sodium hydroxide solution. A blue precipitate forms if magnesium is present. Ensure that each pupil understands the tests fully before going on to test the solids from the sea water.

Experiment sheet 22

The sea contains very large quantities of chemicals – not only common salt but also many other substances, and even gold. There is, of course, also a great deal of water so we must separate the chemicals from this before trying to find out what they are.

You are going to look at and test the substances that separate when 1 dm^3 of sea water is evaporated. The evaporation will have been done beforehand, to save time, and you will be given samples of the solids which separate when the volume has been reduced to 125 cm^3, 25 cm^3, 10 cm^3, and 5 cm^3 – four samples in all. You will need to become familiar with the tests that you are going to use – part (1) below will enable you to do this.

1*a*. You should already know how to test for a carbonate. If you are doubtful look back to your notes on *Experiment* 21.

b. Some metals can be identified by the colours that their

Solid magnesium sulphate

The four solids from the sea water evaporation

compounds give when placed in a Bunsen flame. The test is carried out as follows.

Clean the end of a length of nichrome wire by heating it in the edge of a medium Bunsen flame (non-luminous), dipping it into a little concentrated hydrochloric acid in a small test-tube, and heating again. Repeat this until the wire does not colour the flame. Dip the wire into the acid again, then into a little of the solid to be tested, so that some of the solid is picked up on the wire, and then put the end containing the solid into the side of the Bunsen flame. Note the colour which the substance gives to the flame. Clean the wire, as above, before testing a new substance.

Carry out the flame test with the samples of sodium chloride, potassium chloride, magnesium chloride, and calcium chloride provided. Look at the potassium flame through blue glass also. Record the flame colours in the table below.

Metal in compound	Flame colour
Sodium	
Potassium (viewed through blue glass)	
Magnesium	
Calcium	

c. Prepare separate solutions of sodium chloride, potassium chloride, magnesium chloride, calcium chloride, sodium sulphate, and magnesium sulphate by dissolving small portions of each in about 1 cm^3 water in separate test-tubes. To each tube add an equal volume of silver nitrate solution, acidified with nitric acid. Record what happens in the table below.

d. Wash the tubes thoroughly, giving a final washing with distilled water, and repeat the tests in (*c*) using barium chloride solution acidified with hydrochloric acid instead of silver nitrate solution acidified with nitric acid. Record the results in the table.

How would you test a substance to find whether it is a chloride?

How would you test a substance to find whether it is a sulphate?

Solution	Reaction with acidified silver nitrate solution	Reaction with acidified barium chloride solution
Sodium chloride		
Potassium chloride		
Magnesium chloride		
Calcium chloride		
Sodium sulphate		
Magnesium sulphate		

Experiment sheet 22 (continued)

e. You may still find it difficult to detect magnesium compounds. Your teacher will explain how to do this. State what you do here.

2. Test each of the samples of solids from the evaporation of sea water for the presence of metals, carbonate, chloride, and sulphate. Summarize the results obtained in the table below.

Composition of solids from sea water

Evaporation range	Solid contains
1 dm^3 to 125 cm^3	
125 cm^3 to 25 cm^3	
25 cm^3 to 10 cm^3	
10 cm^3 to 5 cm^3	

When all the tests have been completed discuss the findings with the class. Point out the evolution of gas from the solution in the tests on the first solid, and remind the pupils of the carbonate test. The first solid should contain calcium carbonate and calcium sulphate. The second solid contains sodium chloride, the third should give positive tests for magnesium, sulphate, and chloride, and the fourth contains potassium and magnesium chlorides. This description refers to sea water from the south coast of England.

Suggestion for homework

Atlantic sea water contains about 2 mg of gold per tonne. Invent a method for getting gold out of sea water. Say why you think your method has not been used!

Summary

Pupils should now know some of the dissolved substances in sea water, and be familiar with tests for a number of materials, including carbonates, chlorides, sulphates, calcium, magnesium, sodium, and potassium.

A10.2
What is the effect of electrolysing sea water?

In this section the pupils will see that the electrolysis of sea water yields hydrogen and a gas smelling of swimming baths (chlorine). The electrolysis of the concentrated residue from A10.1 is seen to give a brown vapour (bromine) and hydrogen.

A suggested approach

Objectives for pupils

1. Familiarity with a technique for collecting gases from electrolysis

The pupils already know that electricity can be used to obtain chemicals from solution. They used a battery to extract copper from copper sulphate solution in A1.6. Remind them of this and tell them to try a similar experiment on sea water. Two gases appear, one at each electrode. They will recognize the smell of chlorine as 'smelling like swimming baths'. They may be told that this is chlorine and

2. Knowledge of chlorine and bromine as elements present in sea water

be introduced to the starch–iodide test. They should identify the other gas as hydrogen as they met it in Topic A7. Details of the experiment are given below.

Experiment A10.2a

Electrolysing sea water

Apparatus

Each pupil (or pair) will need:

Electrolysis cell as illustrated, and described in *Collected experiments*, Experiment E5.3, with 2 rimless test-tubes, 75 × 10 mm

2 lengths of connecting wire with crocodile clips

6 V battery or alternative d.c. supply

Splints

Stand and clamp

Full-range Indicator

Starch–iodide paper

Supply of sea water

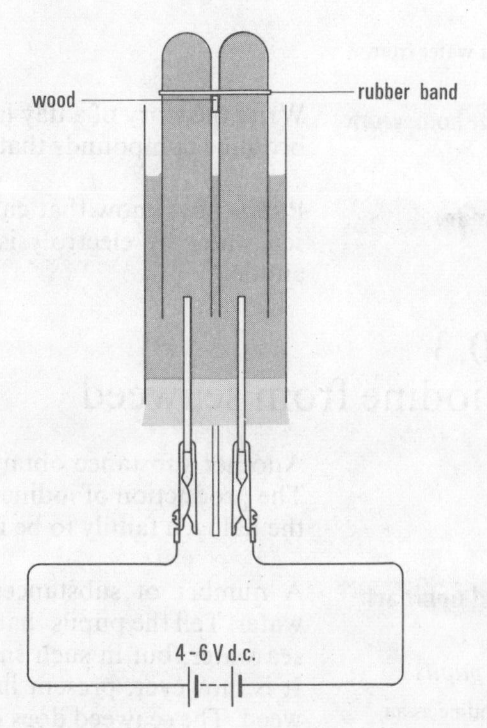

Procedure

The cell is half filled with sea water and the d.c. supply connected. No bulb is needed in the circuit for this experiment. Chlorine will appear at the anode and hydrogen at the cathode. When the gases appear suggest that they should be collected and their properties investigated. The starch–iodide test for chlorine is carried out by holding a piece of moist starch–iodide paper in the gas. Iodine is liberated and forms a deep blue colour with the starch; the paper therefore turns from white to deep blue. The gas liberated at the cathode may be tested using Full-range Indicator paper and a lighted splint. It will be found to burn.

After this experiment by the pupils the experiment with concentrated sea water may be demonstrated. There will probably not be enough of this solution for the pupils to do this themselves.

Experiment A10.2b

Electrolysing concentrated sea water

This experiment should be done by the teacher.

Apparatus

The teacher will need:

2 carbon or steel electrodes

(Continued)

Procedure

The teacher can electrolyse the small amount of residual solution kept from Experiment A10.1. This is insufficient to place in the pupil's apparatus, but a pair of electrodes can be dipped into the

2 lengths of connecting wire with crocodile clips

6 V battery or alternative d.c. supply

Watch-glass

Teat pipette

Concentrated sea water from Experiment A10.1

small amount of solution available on a watch-glass. A yellow–brown colour will be seen to develop around the anode (bromine).

Suggestion for homework

Write the story of a day in your life mentioning all the chlorine and bromine compounds that you come across directly or indirectly.

Summary

Pupils now know that chlorine and bromine can be obtained from sea water by electrolysis, and that these elements appear at the anode.

A10.3 Getting iodine from seaweed

Another substance obtainable indirectly from sea water is iodine. The production of iodine from seaweed allows the third member of the halogen family to be introduced.

A suggested approach

Objective for pupils

Knowledge of iodine as an element present in sea water

A number of substances have already been extracted from sea water. Tell the pupils that another important substance is present in sea water, but in such small quantities that it is difficult to detect. It is, however, present in larger quantities in certain types of seaweed. The seaweed does the concentrating in this case. We can find out what this substance is by drying seaweed and carrying out an extraction.

Experiment A10.3

Apparatus

Each pupil (or pair) will need:

Experiment sheet 23

Beaker, 100 cm^3

Test-tube, 150×25 mm

Bunsen burner, tripod, gauze, and asbestos square

Filter funnel and filter paper

About 1 g of dried seaweed (*Laminaria*)

20-volume hydrogen peroxide

M sulphuric acid

Tetrachloromethane

Distilled water

Iodine from seaweed

Procedure

In *Experiment sheet* 23, pupils are asked to plan an experiment to confirm that iodine has been extracted into the tetrachloromethane layer. There is not sufficient iodine in the layer to form crystals if the tetrachloromethane is evaporated off, but the class results could be pooled in order to attempt the crystallization, which should be done in a fume cupboard.

Experiment sheet 23

Iodine is an important substance which is present in sea water but in such small proportions (about 1 part in 20 million) that it is difficult to detect. It is present in seaweed in much larger proportions. Try to extract a little iodine from seaweed by the following method.

Gently boil about 1 g dried seaweed with about 10 cm^3 distilled water in a beaker for a few minutes. Filter into a test-tube and add about 2 cm^3 dilute sulphuric acid and

10 cm^3 hydrogen peroxide solution to the filtrate.
What happens?

Add about 3 cm^3 tetrachloromethane, shake, and allow the tube to stand for a few minutes. What do you see now? Where do you think the iodine is?

Plan an experiment to check your answer and try it.
What did you do and what happened?

Suggestion for homework

You are given a beaker full of crystals of salts collected from the shores of the Dead Sea. Describe how you would find out what salts are present.

Summary

Pupils now know that seaweed of certain types is able to concentrate iodine present in sea water, and that this iodine can be extracted.

A10.4
A family of elements – the halogens

In this section the three elements chlorine, bromine, and iodine are introduced as a family, but detailed investigation of their properties is left until Stage II. The elements themselves are looked at, together with their colour in aqueous and tetrachloromethane solutions. A reactivity table is then established.

A suggested approach

Begin by showing the pupils some of the physical properties of the halogens. This may be restricted to their appearance, and that of some of their solutions; the following experiment indicates the scope of what is intended.

Objective for pupils
Knowledge of chlorine, bromine, and iodine; their physical characteristics and relative reactivity

Experiment A10.4a

Investigating the halogens
This experiment must be carried out by the teacher, preferably in a fume cupboard.

Apparatus
The teacher will need:

Filter flask, fitted with delivery tube, bung, and tap funnel, with which to generate chlorine

6 stoppered bottles, about 50 cm^3 size

3 test-tubes, 150 × 25 mm, in rack

Concentrated hydrochloric acid

(Continued)

Procedure
Begin by filling a large test-tube with chlorine so that the pupils can see that it is a green gas. Do this by dropping two or three drops of concentrated hydrochloric acid from a tap funnel onto well-ground crystals of potassium permanganate contained in a filter flask (see diagram). Take care not to inhale the gas, and do not let the acid drop too rapidly onto the permanganate. Pour out a little bromine (a red liquid – **care** – very corrosive, wear protective gloves) into a test-tube standing in a rack, to avoid spilling on the hands. Place a few crystals of iodine into a third test-tube. Iodine crystals (seen to be black) produce a deep purple vapour when gently heated.

Potassium permanganate

Bromine

Iodine

Tetrachloromethane

Distilled water

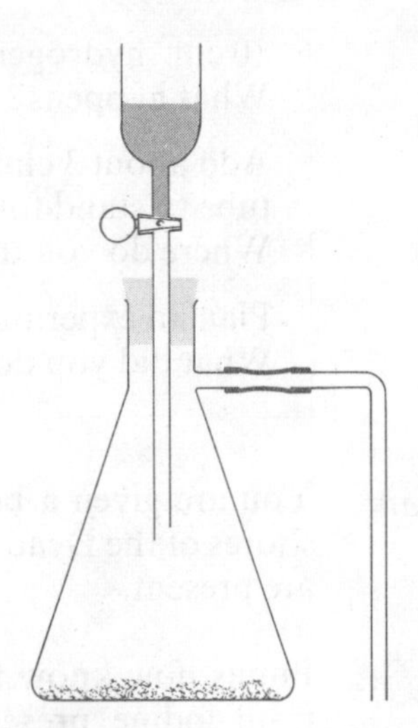

Put 5 cm^3 of tetrachloromethane in each of the three bottles. Bubble some chlorine into the first one, place a few drops of bromine into the second, and a few crystals of iodine into the third. Stopper each bottle securely and shake the contents. Show the pupils the colours of the solutions that are formed (green, red, and purple respectively). Keep these solutions for the next experiment. Repeat the whole experiment with water instead of tetrachloromethane. It will be seen that the halogens are much less soluble in water, and that iodine gives a brown solution.

After showing the pupils the halogens in this way, let them do the next experiment, in which an order of reactivity will be established.

Experiment A10.4b

Competition among the halogens

Apparatus

Each pupil (or pair) will need:

Experiment sheet 24

3 test-tubes in rack

Access to solutions in tetrachloromethane of chlorine, bromine, and iodine (made by diluting the solutions from Experiment A10.4a to 50 cm^3 with more tetrachloromethane)

Sodium chloride solution

Potassium or sodium bromide solution

Potassium or sodium iodide solution

Procedure

This is described in *Experiment sheet* 24 reproduced below.

Experiment sheet 24

You have examined earlier the competition among metallic elements for oxygen and established an 'activity series'. Similar competition can be studied among the non-metallic elements such as the halogens (chlorine, bromine, and iodine). In this experiment you will try to find their order of reactivity towards potassium or sodium.

You are provided with solutions of potassium (or sodium) chloride, bromide, and iodide, together with solutions of chlorine, bromine, and iodine, in tetrachloromethane. Examine the effect of adding each of the last three solutions to each of the first three solutions separately, shaking each mixture and allowing it to stand.

What is the order of reactivity of chlorine, bromine, and iodine? (Put the most reactive first.)

What are your reasons?

Suggestions for homework

1. Read the Study sheet *The halogens*.
2. Make a summary of the work on halogens that you have done in this Topic.

Summary

After this section the pupils will have some familiarity with the halogens chlorine, bromine, and iodine, and should know their 'order of reactivity'.

A10.5
The industrial production of the halogens

A suggested approach

Finally in this Topic the methods of production and the main uses of the halogens should be discussed. Film loops are available to illustrate the modern methods used, and they indicate the depth of treatment which is intended at this stage.

It will be necessary to introduce the remaining halogen, fluorine, as this has so far not been mentioned in this Topic; point out that this extremely reactive gas is not safe to work with in the laboratory.

Objective

Knowledge of the method of production and the main uses of the halogens

The teacher will need

Film loop projector

Film loops:

1–11 'Fluorine manufacture (with notes)

1–12 'Fluorine compounds uses' (with notes)

1–13 'Chlorine manufacture' (with notes)

1–14 'Chlorine uses' (with notes)

1–15 'Bromine manufacture' (with notes)

1–16 'Bromine uses' (with notes)

1–17 'Iodine manufacture' (with notes)

1–18 'Iodine uses' (with notes)

Suggestion for homework

Make a summary of the methods of manufacture and the uses of the halogens.

Summary

Pupils have seen something of the methods of production and the uses of the halogens.

Part 2 Alternative B

Topic B1

Separating pure substances from common materials

Purposes of the Topic

The main purpose of this Topic is to show the pupils what a chemist means by a pure substance. The pupil starting this course will have little idea of what is meant by a pure substance, other than the expression he has seen in the shops, 'guaranteed pure'. During this Topic he should learn:

1. The meaning of the terms 'pure substance' and 'mixture'. The understanding of the concept of a pure substance will continue to grow throughout Stage I, and indeed well beyond this.
2. Something about certain simple techniques for separating substances, notably paper chromatography, distillation, filtration, evaporation, and crystallization.
3. The meanings of a number of new words by experiencing the need to name something rather than by straight definition. Lists of new words are given at the end of each section in this Topic.
4. How the experiments in the laboratory have their counterparts in the practical world of industry, and the importance of the applications of chemistry to everyday life.

Contents

Timing

It is important to give enough time for this Topic.

Sections B1.1, B1.2, and B1.5 will probably need two double periods each, but sections B1.3 and B1.4 should need only one. A total of ten periods for this Topic allows plenty of time for discussion and for following up pupils' own ideas. If time is short, B1.2 could be omitted and B1.5 cut down. In this way the Topic could perhaps be covered in five double periods, but great care is needed to make sure that the pupils have time to absorb all that is new to them.

Introduction to the Topic

After some preliminary work with a Bunsen burner, the pupils try to separate 'pure' substances from impure ones. They separate pure alum from an impure sample by dissolving the impure alum in water, then filtering and crystallizing it; and pure naphthalene from an impure sample using the same technique but a non-aqueous solvent.

Once the ideas of solution, and the recovery of the solute, have been introduced, the difference between tap water and distilled water is investigated. Samples of both are evaporated and the solid content of the tap water discovered. This leads to the question 'How can sea water be purified?' and the answer introduces the technique of distillation. The Topic ends with the use of chromatography to separate chlorophyll from grass.

Alternative approach

A different treatment of the same theme is given in Alternative A, Topic A1.

Subsequent development

The experimental techniques learned in this Topic are of value throughout the pupils' later experimental work, and time spent here is well worth while. These techniques include filtration, evaporation, crystallization, distillation, and paper chromatography. Distillation in particular is taken further in Topic B3, when it is applied to crude oil and liquid air.

The knowledge of what is meant by a pure substance is basic to the rest of the course, and the differences between mixtures and compounds are brought out throughout the rest of the two years' work.

Section B1.5 'What coloured substances can we get from grass?' leads on to a study of the use of coloured substances from plants as indicators in section B2.1.

Further references
for the teacher

Other experiments illustrating the same theme can be found in *Collected experiments*, Chapter 2 'Isolation and recognition of a single substance'.

A general discussion of the investigational method, on which most of the work in the Nuffield Chemistry scheme is based, will be found in Chapter 1 of the *Handbook for teachers*. A general discussion on approaches and attitudes to starting a chemistry course is given in Chapter 3 of the same book.

Physics
Teachers whose classes are studying the Nuffield Physics course should read the Physics *Teachers' guide I*.

Supplementary materials

Film loops
1–1 'Salt production'
1–2 'Chlorophyll extraction'

Reading material
for the pupil

Study sheets:
Analysis. In this, pupils are helped, by pictures and words, to understand what analysis means to a chemist. They are also shown how analysts can help us by checking samples of food and air, for example, and how forensic scientists can provide vital clues in criminal investigation.

Water is broad in approach, to expand pupils' horizons on this substance which they see so often both inside and outside the laboratory. It deals with how water affects the lives of people all over the world, and with ways of ensuring that we have enough water in the future.

The words chemists use contains a glossary of expressions used in Stage I, a short account of the derivation of various words, and a table of dates and discoverers of some elements. It can be used at any part of Stage I but many teachers may like to use it at the beginning as an introduction to the new and, perhaps, strange words that pupils will meet.

Background book:
Growing crystals

Starting the course – the first lessons

There are many ways of starting a chemistry course. One way to capture the pupils' interest and curiosity in the subject is to have on the bench a large number of everyday substances: detergent, a brick, a piece of marble, some paint, a bottle of aspirins, and indeed anything that is relevant and readily available. You can then explain to the class that chemistry is concerned with all these things – finding out about the common materials around us, and then shaping them to our own needs. More is said later about this approach.

During the introductory lesson or lessons the pupils will, of course, have to be shown the laboratory and possibly the science library. A certain amount of administrative work will no doubt also be done so that this introduction may take either one or two double periods.

Before starting the course proper it is worth spending a lesson, or part of a lesson, in discussing and learning to use the Bunsen burner. The pupils will have to use the burner in the first experiment that they do and it is more convenient to deal with it in advance than to have to stop in the middle of an experiment to do so.

The Bunsen burner – a suggested approach

Start by telling the class of the importance of the Bunsen burner in simple laboratory chemistry. It is remarkable that this piece of apparatus, which Bunsen invented for his new laboratory in Heidelberg in 1855, remains virtually unaltered today. Tell the pupils to close the air-hole, turn the gas full on and light it. Now they can all try slowly opening the hole. What do they notice? How many things change when the air-hole is opened? The luminosity of the flame (the question 'why?' can be answered later); the length of the flame; its temperature (ask them if it seems to get hotter); the noise it makes (ask them why). Now get the pupils to open the air-hole fully and gradually turn the gas down until the flame strikes back*. Some will notice that the flame is still burning at the bottom of the burner. Tell them to turn the gas off, close the air-hole (with care as the barrel may be very hot), and relight the burner. They may now examine the cause of the luminosity of the flame by closing the air-hole again and holding a piece of broken porcelain (from, say, a crucible lid) in the flame with a pair of tongs. Ask them to note what happens. Now get them to hold the blackened porcelain in a roaring flame. They will see that the black deposit disappears.

**This may not happen with North Sea gas.*

Another way to open the lesson on the Bunsen burner is to arrange that half the Bunsen air-holes are open and half closed before the lesson starts. Let the pupils light their Bunsen burners and then point out that half of them have a luminous flame and half have a non-luminous flame. Ask them why this is so.

Let the pupils have a closer look at the flame itself. The characteristics of the different parts of the flame are nicely shown by using a piece of asbestos paper. Hold the paper flat in the flame (air-hole open) and see which part of the flame is hot enough to make the asbestos red-hot. Now lift the paper up through the flame and watch the rings decrease in size. Ask them which they think is the hottest part of the flame.

Let the pupils practise setting the burner with the gas *half* on and with the air-hole *half* open. This is the kind of flame which they will usually require and is also the flame we shall mean when we say 'heat with a Bunsen flame'.

B1.1
How can crude alum be purified?

In order to teach the pupils what a pure substance is, and how a substance can be purified by crystallization, they are asked to suggest ways of purifying a sample of crude alum. They devise an experiment and obtain crystals of pure alum from the crude material.

A suggested approach

Objectives for pupils*

1. Skill in the techniques of filtration, evaporation, and crystallization
2. Awareness of what the chemist means by a 'pure' substance
3. Understanding of the meaning of the words: dissolve, solvent, solution, filter, filtrate, evaporate, crystal, crystallize

The pupils have already been introduced to chemistry as a study of materials and have learnt one simple technique, that of using the Bunsen burner. In this section they meet a fundamental concept: 'What a chemist means by a pure substance', and they are given their first problem to solve: 'How can we get pure alum from crude alum?' The initial discussion should centre on the meaning of a pure substance. The pupils will have some idea of this from the everyday meaning of the words 'guaranteed pure'. In this case the concept of a pure substance can be linked with that of a single substance. How do we know if a substance is pure or not? If it is not pure, we can often *see* that there is something else in it. That is to say it is no longer one substance, it has 'impurities' in it. Now take as an example some crude alum. Simply by looking at it the pupils will be able to see that it is 'dirty' and that it contains more than one substance. Pass some around the class, and present the first problem: 'How can we purify this alum? How can we get rid of the dirt?' They may suggest putting it in water to wash it off.

If they do, try it yourself or let them try it. They will see that the crystals dissolve but the 'dirt' remains behind. The idea of dissolving may already be familiar to them through everyday experiences like dissolving sugar in tea. The next problem is how to get rid of the dirt.

Someone may suggest 'sieving it off'. If so, this gives a lead for the introduction of filtration. The use of 'quick' filter paper, for example, Whatman No. 41 or Green No. 802, saves time here. The liquid which comes through is clear and the solid can be seen on the filter paper. Where is the alum? It is dissolved in the water and we have a solution of alum in water. The last problem is 'How are we to get the alum back from the solution?' Someone may suggest driving off the water by boiling. Let the pupils try this and find out for themselves that there is a point at which crystals begin to appear.

*For the sense in which this word is used, see Part I, page 13.

The last point for discussion in this section is the use of crystallization as a means of purification. Simply driving off the water does not purify the alum, but allows the crystals to form as the solution cools. The impurities then remain in the 'mother liquor' surrounding the crystals, which may then be separated from the 'mother liquor' by filtration. The technique for these experiments is described below.

Experiment B1.1

How can crude alum be purified?

Apparatus

Each pupil (or pair) will need:

Experiment sheet 25

Beaker, 100 cm^3

Crystallizing dish

Funnel and quick filter paper, e.g. Green No. 802 or Whatman No. 41

Bunsen burner, tripod, gauze, and asbestos square

Glass rod

Stand and clamp or a conical flask for supporting funnel

Hand lens or access to low-power microscope

Sample of crude alum, about 5 g; if none is available, a suitable mixture may be made by adding dried soil or dust to alum crystals

Procedure

This is described in *Experiment sheet* 1 from the pupils' book *Experiment sheets I*. To aid the teacher, this sheet is also reproduced below. (All the *Experiment sheets* are reproduced at the relevant points in this book, and are distinguished from the rest of the text by having a line round them.)

While the experiment is being done, pupils could be told that the crystals of alum form on cooling because its solubility in cold water is much less than in hot water. If crystals fail to appear, the pupils should be told to boil off some of the water, and allow the solution to cool in the crystallizing dish once more.

After the experiment you may like to let the pupils look at their crystals under a hand lens or low-power microscope. Some of the pupils may like to 'grow' one of their crystals. This is a good place to initiate a brief and quite informal discussion on orderliness, as exemplified by the alum crystals (and the naphthalene crystals in Experiment B1.2). This will be expanded later into a treatment of atomic and molecular structure. See the *Handbook for teachers*, Chapter 9.

Also see the suggestions on building up a pupil's record of experiments in Appendix 2, page 234, of this book.

Experiment sheet 25

An important task in chemistry is the purification of substances by removal of impurities. In these first experiments in this course you will have an opportunity to try some of the methods that chemists use for this purpose. In this experiment you will try to purify some crude alum. You will have discussed possible ways of doing this. The notes given below are intended as reminders.

Use enough powdered impure alum to cover the bottom of a beaker and add enough water to cover it. Heat the mixture gently with a medium Bunsen flame, supporting the beaker on a wire gauze resting on a stand. Stir the mixture with a glass rod and continue heating until the water boils. If all the alum has not dissolved (neglect the dirt), add a little more water and heat again. Continue in this way until as much of the crude alum as possible has dissolved. Filter the solution while still hot (hold the beaker with a cloth) into a clean dry dish.

Allow the filtered solution to cool. A crop of crystals should appear. Pour the liquid from them into a clean beaker. Dry the crystals on filter paper or similar absorbent material (paper towel or tissue).
Examine a few of the crystals with a hand lens or low-power microscope.
Can you describe their shape?
Draw one of the crystals.

The liquid in the beaker is a saturated solution of alum. You may like to use it for growing larger alum crystals. Ask your teacher about this.

New words

Dissolve
Solvent
Solution
Filter
Filtrate
Evaporate
Crystal
Crystallize

Suggestions for homework

1. How would you find out if soil contained some substances that are soluble in water?
2. Obtain some crystals from ordinary salt using home utensils; for example, an egg cup for solution, a saucer for evaporation. Are the crystals like alum crystals?
3. Read the Background book, *Growing crystals*.
4. A number of words are used in chemistry which are not in common use elsewhere, and a number of everyday words have special meanings in chemistry. Start to collect new words in a 'Chemistry word book' in alphabetical order for easy reference.

Summary

Pupils should now be able to use the techniques of filtration, evaporation, and crystallization. They should understand the meaning of the words dissolve, solvent, solution, filter, filtrate, evaporate, crystal, crystallize; and they should have some awareness of what is meant by a pure substance (namely, a single substance).

B1.2
How can crude naphthalene be purified?

Crude naphthalene is chosen as a second example of a substance to be purified. In this case it does not dissolve in water and a non-aqueous solvent must be tried. The centrifuge is introduced as a means of separating liquids from solids.

A suggested approach

Start the lesson by showing the class a sample of crude 'gas-works' naphthalene beside some moth-balls. The moth-balls represent pure naphthalene which, you can say, is said to have been obtained from

Objectives for pupils

1. Skill in the techniques of centrifuging, filtration, and the handling of a volatile solvent
2. Understanding what the chemist means by a pure substance
3. Understanding of the meaning of the words non-aqueous solvent and centrifuge

Useful BBC programme

'Coal age?'

See Appendix 4 for a brief description and further details

the crude naphthalene. Ask pupils to suggest how the crude naphthalene may be purified. They will probably suggest dissolving it in water, as they did with the alum in the last section. If they do, let them try it. They will find that it melts, but does not dissolve, although some may vaporize in the steam if the water is boiled. Ask them if they have any other suggestions: 'Is there anything else it will dissolve in?' 'What happens when you send clothes for dry-cleaning?' They will have heard of dry cleaning and may know that this is done with a solvent which contains no water. The word solvent may be introduced here. Have some 'dry cleaning liquid' at hand and allow them to try to dissolve a spatula measure of naphthalene in a test-tube with about 2 cm depth of the liquid in it. The solvents recommended are 1,1,1-trichloroethane (which is non-toxic and non-flammable) or ethanol which is inflammable. Carbon tetrachloride should not be used as it is toxic. When the pupils have seen that naphthalene does dissolve in the liquid, an experiment may be devised to purify the crude naphthalene. This may be done quickly and efficiently if the procedure outlined below is followed.

Teachers should note that there are many new points introduced in this section. They include the use of a centrifuge, filtration under reduced pressure, decolorization with activated charcoal, and non-aqueous solvents. It is important therefore to allow plenty of time for each of these points to be properly understood.

Experiment B1.2

How can crude naphthalene be purified?

Apparatus

Each pupil (or pair) will need:

Experiment sheet 26

2 test-tubes, 100 × 16 mm

Beaker, 100 cm^3

Glass rod

Tripod and gauze

Bunsen burner and asbestos square

Buchner or Hirsch funnel, filter paper, filter flask, and filter pump

Teat pipette

Spatula

Filter papers for drying crystals

Thermometer, −10 to +110 °C

(Continued)

Procedure

Making the solution:

1. *Using 1,1,1-trichloroethane as solvent.* Pupils should put about 2 cm depth of crude naphthalene in a 100 × 16 mm test-tube and add 2 cm^3 of trichloroethane. The test-tube should then be warmed over a small Bunsen flame, stirring the contents with a glass rod to aid solution.
2. *Using ethanol (industrial methylated spirit) as solvent.* **Warning:** *ethanol is inflammable.* Pupils should take a 100 cm^3 beaker, fill it two-thirds with water, and heat it to about 80 °C on a tripod and gauze. They should then place the beaker at bench level on an asbestos square and extinguish all Bunsen flames.

About 2 cm depth of crude naphthalene should be put in a 100 × 16 mm test-tube and the tube half filled with ethanol. Place the test-tube in the beaker of hot water, stirring with a glass rod to aid solution. When the solution has been made, the contents of the test-tube should be centrifuged hot, to separate insoluble impurities, and the supernatant liquid poured off as quickly as possible into another test-tube to prevent the naphthalene from beginning to crystallize in the centrifuged test-tube. This second test-tube must be thoroughly cooled until a good crop of naphthalene crystals has been obtained. These crystals should then be separated by filtering them through a Buchner funnel and washing them with a few drops

Crude naphthalene. (This is available from laboratory suppliers. Dirty naphthalene can be made by adding a little soil or laboratory dust to a quantity of the clean compound)

1,1,1-Trichloroethane or ethanol (industrial methylated spirit)

Activated charcoal powder

Pupils will require access to a centrifuge

of cold trichloroethane or ethanol. Finally dry the crystals by placing them on a few pieces of filter paper.

If a discoloured sample of naphthalene is obtained, as is most probable with the industrial crude naphthalene, repeat the entire procedure, using a fresh sample of crude naphthalene, but this time adding a spatula measure of activated charcoal to the solution just before warming the solvent and crude naphthalene. A white sample of naphthalene will be obtained this time.

Experiment sheet 26 (reproduced below) refers only to the use of alcohol as solvent, and it is left to the teacher to explain how to use a centrifuge, and how to filter under reduced pressure.

Experiment sheet 26

Naphthalene is one of the products of heating coal. It is very impure (a nasty-smelling black substance) when first obtained. You will have discovered that it does not dissolve in water which means that you must try a different liquid as solvent.

Put about 2 cm depth of crude naphthalene in a test-tube.

Put tap water into a beaker until it is about two-thirds full and heat it on a gauze supported on a stand until the temperature of the water is about 80 °C. *Turn out the Bunsen burner* – the liquid you are going to use is inflammable. Using a cloth to hold it, stand the beaker of hot water on an asbestos mat on the bench. Add ethanol (alcohol) to the crude naphthalene until the test-tube is half full. Put the test-tube in the hot water and stir the contents until most of the naphthalene has dissolved.

Remove the test-tube from the hot water, centrifuge the contents at once, and pour off the clear liquid into another test-tube. Cool the outside of this tube in a stream of water from the tap. What happens?

Separate the crystals by filtering (your teacher will tell you how to do this), wash them with a few drops of ethanol, and dry on absorbent paper. Describe the appearance of the crystals.

Now repeat the whole process (re-heat the water if necessary but remember to put out the flame before bringing ethanol near the beaker) but add a measure* of powdered charcoal to the mixture of crude naphthalene and ethanol in the test-tube used at the beginning of the experiment.

How does the naphthalene obtained this time differ from that you produced in the first purification?

*A 'measure' means the amount of solid that would lie on a new penny, or would just fill the grooved end of a Nuffield spatula.

After the experiment discuss the results with the class. Emphasize the fact that a solvent other than water was used in this case and discuss the use of solvents in industry.

New words

Non-aqueous solvent
Centrifuge

Suggestions for homework

1. Read the Study sheet *Analysis*.
2. Why do you think 'dry cleaning' is so called? What does the dry cleaning solvent do that water cannot do?
3. Imagine that you have just bought a bottle of a crystalline chemical which is soluble in water. Unfortunately, the bottle is knocked off the shelf and breaks. Explain how you would obtain clean crystals from the mixture of crystals, dust, and glass swept up from the floor.

Summary

Pupils should now be able to use the techniques of centrifuging, and filtering under reduced pressure, and be familiar with the use of non-aqueous solvents.

B1.3
What is the difference between tap water and distilled water?

The question 'What is the difference between tap water and distilled water?' is used as a basis for introducing the idea of purification by distillation.

A suggested approach

Objectives for pupils

1. Skill in the technique of evaporation over a water bath
2. Use of the idea of a 'pure' substance – that is, a single substance
3. Awareness of the nature of distilled water
4. Recognition of the process of distillation

Ask the pupils if there is any difference, for instance, between rain water, sea water, and tap water. They will certainly know that sea water tastes salty and some of them will have noticed that rain water tastes different from tap water. River water is dirty-looking stuff (in some rivers at least). Can we purify it just as naphthalene was purified? There are two sorts of water in the laboratory: tap water and distilled, or deionized, water. Ask them if they can think of any way of finding out whether there is a difference. They both *look* the same. Suggest that they try evaporating the water to see if anything is left behind. This can be done on a microscope slide under an infra-red lamp or using a beaker of water and watch-glass as described below. In either case, the tap water leaves a residue behind. This means that there was something other than water in the tap water. Thus tap water is not a pure substance. Now let the pupils try the experiment of which the details are given below. (Make sure the tap water is hard!)

Experiment B1.3

Investigating different kinds of water

Procedure
This is described in *Experiment sheet* 27, reproduced below. The water in the beaker should be boiled fairly briskly so that the steam

Apparatus

Each pupil (or pair) will need:

Experiment sheet 27

Watch-glass (hard-glass), approximately 75 mm diameter

Beaker, 100 cm^3

Bunsen burner, tripod, gauze, and asbestos square

A cloth should be available

Alternatively

Microscope slide for each pupil

Infra-red lamp for class use

Access to samples of distilled water and tap water; also sea water, river water, etc., as available

produced causes the water on the watch-glass to evaporate. As soon as the water on the watch-glass has all gone (about ten minutes will be required), remove the burner and take the watch-glass off the beaker, drying the drops of water adhering to the underside with a cloth. A residue will be seen on the watch-glass.

Do not heat the watch-glass directly with a burner; it may crack.

If the evaporation is done under an infra-red lamp, drops of water may be placed on a microscope slide. (They will not run off the slide if it is first made slightly greasy by wiping with the finger.)

Experiment sheet 27

You will know of several different sorts of water – tap water, river water, rain water, pond water, distilled water, sea water are examples. Are they really the same or are there differences? One possible way of answering this question is to evaporate samples of the different sorts and see what is left behind.

You can either evaporate the same number of drops (six would be convenient) of different kinds of water on separate clock glasses heated by steam over boiling water in a beaker, or heat single drops on a microscope slide under a heating lamp. (You can get three single drops on a slide if you are careful – make it slightly greasy first by rubbing it with a finger – but you must make a careful note of which is which.)
How do the residues obtained compare with each other?
Which sample of water contained the most dissolved solids?
Which contained the least?

A brief discussion of the planned energy transfers involved in evaporation (and later in distillation) could be usefully initiated here. See Chapter 15 of the *Handbook for teachers*.

After the pupils have completed this experiment satisfactorily, ask them if it would be possible to get back the water they have driven off. What would we have to do? They will make suggestions such as 'cool it down again', and out of these you can build up the idea of distillation. This will be explored more fully in the next section.

Suggestions for homework

1. If you were in a tropical country by the sea with no fresh water available, how would you try to get pure water from sea water?
2. Rivers flow down to the sea and yet river water does not taste as salty as sea water. How do you account for this? How could you test your explanation?

Summary

Pupils should now be able to evaporate liquids over a water bath. They should be aware that distilled water contains no dissolved solids whereas most other waters do contain solids in solution, and they should have some idea of what distillation involves.

B1.4
How can water from the sea be purified?

On the basis of experience gained in the last section pure water is obtained from sea water by distillation.

A suggested approach

Objectives for pupils

1. Skill in the technique of simple distillation
2. Knowledge of the use of boiling points to identify liquids
3. Understanding what the chemist means by a 'pure' substance
4. Knowledge of the process of salt production
5. Understanding of the meaning of the words: distil, distillation, boiling point, condense, condenser, thermometer

The teacher will need

Film loop projector

Film loop 1–1 'Salt production' (with notes)

In the last section the pupils, by evaporating samples of distilled water and tap water, found that the main difference between the samples of water was in the dissolved solids in the tap water. They were left with the idea that a process of evaporation and condensation could be used to purify water. If they did the suggested homework, the lesson may begin with a discussion of their answers to the problem of getting pure water from sea water. From the discussion that follows a simple apparatus may be built up, and an experiment based on this apparatus is described below.

Experiment B1.4a

Apparatus

Each pupil (or pair) will need:

Conical flask, 100 cm³

Cork or bung with one hole, to fit the flask

Length of glass tubing bent as in the diagram on *Experiment sheet* 2, page 41

Test-tube, 100 × 16 mm

Measuring cylinder, 25 cm³

Stand and 2 clamps

Tripod stand and gauze

Bunsen burner and asbestos square

Supply of sea water, approximately 10 cm³

How can we obtain pure water from sea water?

Procedure

There is no *Experiment sheet* for this experiment. A suitable apparatus, and instructions for assembling it, are given on *Experiment sheet* 2 (Experiment A1.2a). Pupils should be warned to stop heating before the flask becomes dry.

After the pupils have each obtained a little liquid, stop the experiment and discuss the result with them. Ask them how they would find whether the liquid in the test-tube is pure water. They may well suggest a repeat of Experiment B1.3 (*Experiment sheet* 27) to answer this question, in which case they should be encouraged to do so. As a further identification, the teacher should then introduce the idea that each liquid has its own boiling point and suggest this as a means of identifying pure water. To find the boiling point of the liquid, they will need more than a few drops and a more efficient apparatus for condensing it. Discuss this problem with the class and produce the Liebig condenser as an answer to the problem. The following experiment may now be demonstrated.

Experiment B1.4b

An efficient way of purifying sea water

This experiment should be done by the teacher.

Apparatus

The teacher will need:

Distillation flask (at least 100 cm^3 capacity)

Liebig condenser and connection tubing to tap and water

Thermometer, -10 to $+110\,°C$

Beaker, 100 cm^3

2 stands and clamps

Tripod, gauze, Bunsen burner, and asbestos square

Corks or bungs to assemble apparatus*

Anti-bumping granules

Supply of sea water

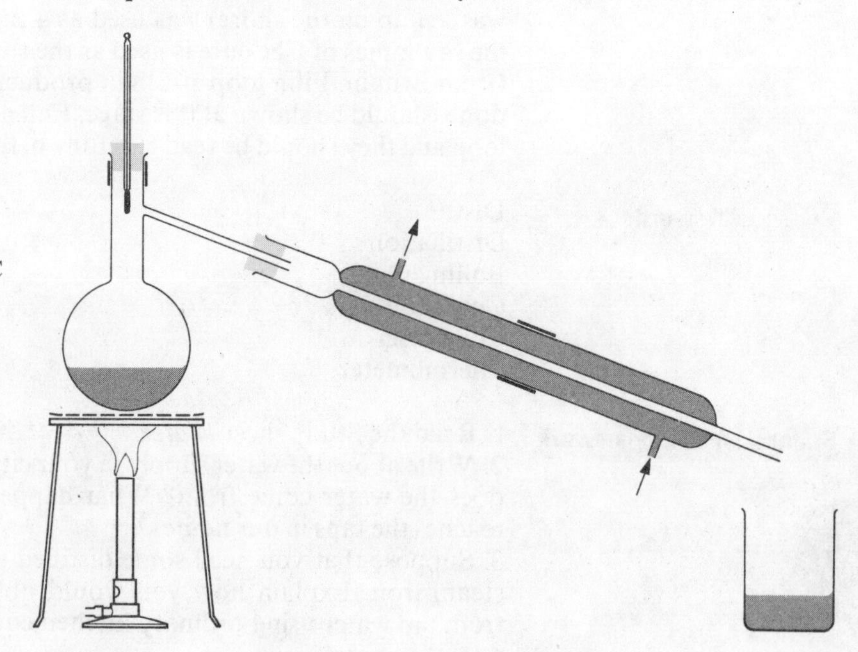

Procedure

The distillation apparatus should be set up as shown in the diagram. Important points to note are:

1. The thermometer bulb should be just below the side-arm of the distillation flask to record the temperature of the vapour actually being collected.
2. The cooling water should enter through the lower inlet and leave by the upper inlet so as to keep the condenser jacket full of cooling water.
3. A few anti-bumping granules should be placed in the distillation flask to ensure even boiling. (There is an opening here for a discussion on the reason why.)

Collect sufficient water for all the pupils to see, but do not boil the contents of the flask dry, or it may crack.

During this experiment read the thermometer to determine whether the liquid being distilled has the same boiling point as that of water. Water from the tap may then be distilled in the same apparatus to see if the same temperature reading is obtained.

Before finishing this Topic some mention of industrial applications may be made. Large-scale evaporation of sea water is carried out in countries such as Israel where fresh water is scarce and the climate is hot. The processes used are not as simple as the one the pupils have used, but the object is the same. The evaporation of sea water to

*Corks or bungs will not be needed if apparatus fitted with interchangeable ground-glass joints (for example, Quickfit apparatus) is used. Although more expensive, this type of apparatus is much more convenient, and is recommended for the teacher's own use.

obtain salt used to be carried out in this country in salt pans such as those once found near Blyth in Northumberland. Sea coal (coal washed up on the shore) was used as a fuel. Nowadays brine from the salt mines of Cheshire is used as the source of most of the salt in Great Britain. Film loop 1–1 'Salt production', showing how this is done, should be shown at this stage. Full notes are supplied with the loop and they should be read carefully before the film loop is shown.

New words

Distil
Distillation
Boiling point
Condense
Condenser
Thermometer

Suggestions for homework

1. Read the Study sheet *Water*.
2. Write about the water supply in your city, town, or village. Where does the water come from? What happens to the water before it reaches the taps in our homes?
3. Suppose that you need some distilled water for a car battery or steam iron. Explain how you would obtain some distilled water from tap water using ordinary kitchen equipment.

Summary

After this section pupils should have some skill in the technique of simple distillation. They should know that boiling points are useful to identify liquids, and that distillation is a good way to purify sea water; and they should have some knowledge of the way in which salt is produced.

B1.5
What coloured substances can we get from grass?

In this section the pupils follow the search for pure substances by trying first to extract the 'green stuff' from grass and then seeing if this 'green stuff' itself is pure.

A suggested approach

Objectives for pupils
1. Ability to design an experiment
2. Skill in the technique of paper chromatography
3. Use of the chemist's idea of 'pure' substance
4. Knowledge of the industrial process of chlorophyll extraction
(Continued)

Is grass* green because it is made entirely of one green substance or is there a green stuff which colours it? Some pupils may recall getting green stains on their clothes, and this may make them think that the green colour comes out of the grass. In any case, ask the pupils how to find the answer to this question: 'Is it possible to dissolve the green out of the grass in the same way that you got the pure alum from crude alum?' They will soon come to the conclusion that, if this were the case, the grass would have its colour washed away by the rain. But what if the grass is crushed? Get the class to try crushing a handful of cut grass with a little water in a pestle and mortar. They may find that after some time the water becomes green. If it does, get them to filter the 'green water'. The green colour remains on the filter paper. This experiment has (in one sense) failed. Ask the

*Fresh spinach is a good substitute for grass.

5. Understanding of the meaning of the words: chromatography and chromatogram

The teacher will need

Film loop projector

Film loop 1–2 'Chlorophyll extraction' (with notes)

class for more ideas. 'If the green stuff (assuming there is such a substance) does not dissolve in water, will it dissolve in anything else? If water does not get the grass stain out of clothes, what does?' Remind the pupils of the solvents they used in B1.2 and show them another: acetone. Now get them to try crushing the grass with a little of this solvent. This time they will find that a green solution is formed and that it will pass through a filter paper. The green solution may now be tested to see if it contains only one green substance or more than one substance. The method used for this is paper chromatography, a method which may be introduced by referring to the fact that an ink drop on a piece of blotting paper will sometimes separate into different colours. To get good results with the paper chromatographic separation of the extract of grass you need to pay careful attention to practical details. Details are given below.

Experiment B1.5

An attempted separation of the green colouring matter in plants by chromatography

Apparatus

Each pupil (or pair) will need:

Experiment sheet 4

Pestle and mortar

Test-tube, 100 × 16 mm

Beaker or evaporating basin as support for filter paper

Filter paper (preferably Whatman No. 1)

Teat pipette

Scissors (unless the grass is already cut up)

Supply of grass (or spinach)

Acetone

Toluene

Procedure

This is described in *Experiment sheet* 4 reproduced below.

Experiment sheet 4

You have seen that some kinds of ink contain mixtures of coloured substances. In this experiment the same method will be used to find out if the green colour of plants is due to a single substance or a mixture of substances.

Take a small handful of grass or leaves and cut it into small pieces. Put the pieces into a mortar, add about 3 to 4 cm^3 of acetone and grind the mixture with a pestle. Add a little more acetone if it seems to be needed. Pour off the liquid into a test-tube.

Has any colouring matter been extracted by the acetone?
What colour is the liquid?

Rest a filter paper on the top of a beaker or evaporating basin. Put one drop of the liquid obtained on to the centre of the paper. Allow it to dry, add another drop and allow this to dry. Repeat until four drops have been added. Now add separate drops of pure acetone, or of toluene, allowing each one to spread before adding the next.
What happens?
Is the green colouring matter in grass a single substance or a mixture?

Allow the filter paper to dry (acetone and toluene are both inflammable).
Draw a diagram of the chromatogram obtained.

If you have time you can try a rather different method of making a chromatogram, using the apparatus shown [overleaf]. Use four drops of the coloured acetone solution as before, and toluene as the developing liquid.

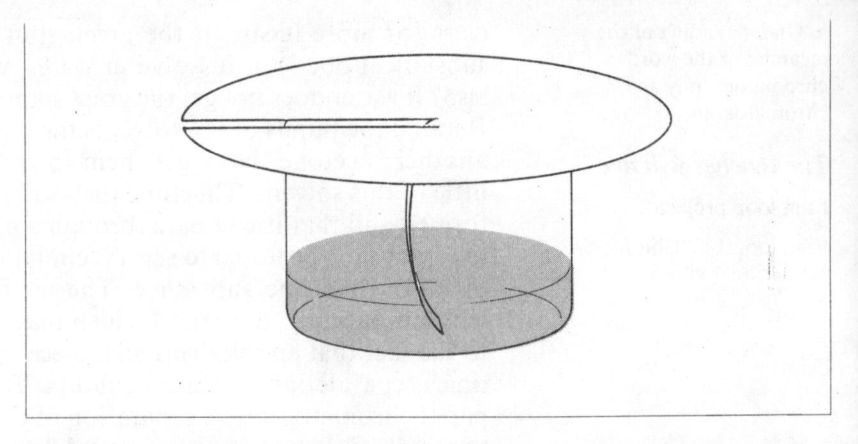

In the chromatogram the outer orange band is xanthophyll and the green band is chlorophyll. There are two types of chlorophyll (chlorophyll a and b), but these are not separated by this technique. Carotene is also present. An inner ring of carotene can usually be seen if toluene is used as the eluent.

The pupils may take some time to learn how to make good chromatograms, but the time is well spent. Let the pupils keep the chromatograms to stick in their laboratory note books.

When the experiment is over, show film loop 1–2 'Chlorophyll extraction'. This loop shows how chlorophyll, xanthophyll, and carotene are extracted from grass on a large scale. Detailed instructions on how to use the loop are given in the film loop notes.

Finally, if there is time, dip a piece of blackboard chalk (plain white, not the 'dustless' or yellow-covered variety) into the green solution. The appearance of yellow and green rings will show that another type of chromatography is taking place in the chalk. See *Collected experiments* for a fuller description of this experiment.

New words

Chromatography
Chromatogram
Chlorophyll
Xanthophyll

Suggestion for homework

Take pieces of filter paper home and make chromatograms of any inks and dyes you can find. Record these results in your laboratory note books. If you have a set of felt-tip pens, use chromatography to find out how many dyes the manufacturer used to make up the different colours.

Summary

Pupils should now be familiar with the technique of paper chromatography, and have some ability to select the most useful technique for separating an unknown mixture. They should also have a knowledge of the industrial extraction of chlorophyll.

Topic B2

Acidity and alkalinity

Purposes of the Topic

This Topic introduces pupils to the idea of acidity and alkalinity. The Topic will enable them to recognize acids and alkalis using the pH scale, and to see how an alkali neutralizes an acid.

Contents

Timing

The first section will probably require two double periods and the second, one. Allowing time for discussion and recapitulation, four double periods are recommended as a maximum. If time is short the second section could be omitted and the first kept to two double periods.

Introduction to the Topic

There are two main points made in this Topic. The first is that acidity and alkalinity, which are treated as one phenomenon, can be recognized by means of some naturally occurring coloured substances (indicators). We introduce pH numbers as a means of referring to a particular degree of acidity or alkalinity as shown by an indicator.

The second is that the acidic properties of solutions can be removed by adding substances (most of which produce alkaline solutions with water) to them. The example taken to illustrate the point is the neutralizing of acidic soil with lime. This example is one which is relevant at least to country children; town children may find the acids in the kitchen more familiar.

Some suggestions about the introduction of the idea of an acid, and of subsequent treatment, are given in the *Handbook for teachers*, Chapter 2, pages 29–31.

Subsequent development

The pH scale and the idea of acidity–alkalinity is needed in Topic B4 during the investigation of the products obtained by burning various materials in oxygen. The subject is developed in more detail in Stage II.

B2.1
Ways of detecting acidity

Acidity–alkalinity is introduced by means of familiar household and garden substances. Taste, natural dyes, and Full-range Indicator are used in turn to detect changes of acidity. The pH scale is introduced as a useful means of describing the acidity of a solution.

A suggested approach

This subject may be introduced in a way which is interesting to the pupils by getting them to take home a large test-tube or small bottle with a cork and to bring it back full of some coloured vegetable or fruit juice extract. Many of these colouring matters are suitable and a list of some of them is to be found below. Alternatively, make your own solutions and have a solution of 'extract of lichen', that is, litmus, as well.

Objectives for pupils

1. Ability to identify acids and alkalis using an indicator
2. Use of the pH scale to give a numerical value to degrees of acidity–alkalinity
3. Understanding of the meaning of the words: acid, alkali, neutral, indicator

Experiment B2.1a

To prepare solutions of some coloured substances from natural sources

Apparatus

The teacher will need:

Pestle and mortar

Round-bottom flask, 250 cm^3

Liebig condenser to fit flask

Beaker, large enough to hold flask

Tripod

Bunsen burner and asbestos square

Filter funnel, stand, and papers

Anti-bumping granules

Plant material

Industrial methylated spirit

The following method gives solutions suitable for the pupils to use in this section. Solutions can be made with, for example, delphinium flowers, blackcurrants, rose petals of various colours, red cabbage leaves, blackberries, and beetroot. Almost all coloured plant material is suitable, except yellow flowers such as daffodils and dandelions.

Procedure

First crush the plant material with a pestle and mortar and then grind it up thoroughly with a mixture of equal volumes of industrial methylated spirit and water. You will require a quantity of this mixture at least ten times the weight of the plant material being used.

Transfer the material to a round-bottom flask (which should not be more than half filled), and add two or three anti-bumping granules to ensure even boiling. Insert a condenser vertically in the flask, and reflux the contents gently over a boiling water bath until it can be seen that the solid parts of the mixture have become 'white'. This should take about twenty minutes.

Allow to cool and then filter to obtain the coloured extract.

At the beginning of the lesson the pupils will have their plant extract in front of them. Ask them what they have found. Did it come from the garden? What is the soil like in this garden? Poor? Sour? What do you do to improve the soil? In the country someone will sooner or later suggest lime. It 'cured' the soil – in other words got rid of the sourness. This will mean nothing to town children living in flats. The kitchen is a more familiar place to them. What do they associate with a sour or sharp taste? Vinegar, lemons, acid drops. Has anyone heard that there is acid in your stomach? Have you ever had stomach ache? What is it due to? Too much acid in your stomach. What do you take for it? Stomach powder. It removes the acidity. The effect of stomach powder on an acid substance can be observed by using the coloured substances from plants. This leads to a search for better ways of showing changes in acidity and to the introduction of an acidity–alkalinity indicator.

Photograph opposite

The teacher helping pupils who are preparing coloured solutions from a variety of natural substances (Experiment B2.1a). This experiment lends itself well to collaborative work.

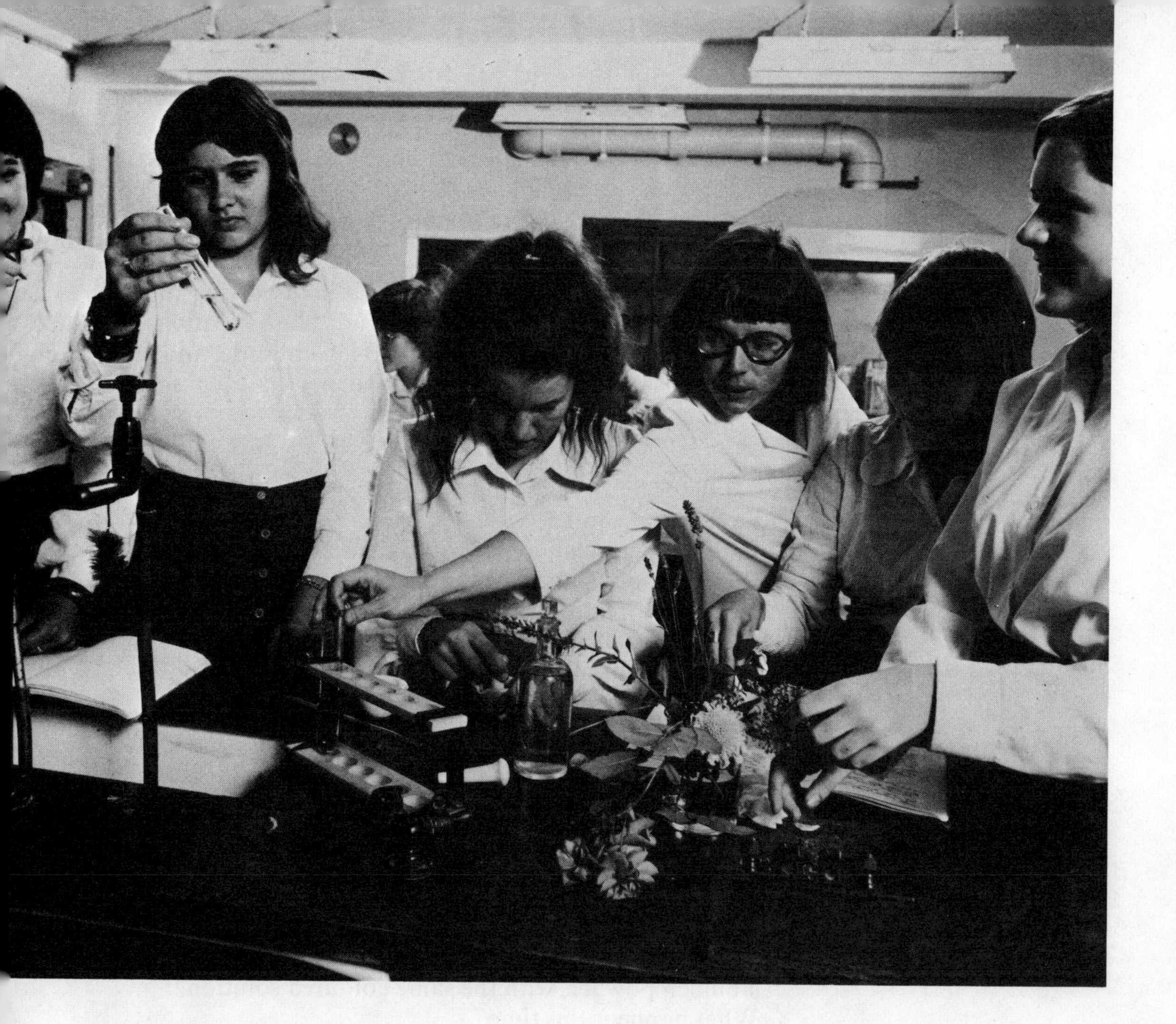

Experiment B2.1b

Using some other coloured substances from plants

Apparatus

Each pupil (or pair) will need:

Experiment sheet 5

6 test-tubes, 100 × 16 mm, and rack

Teat pipette

Supply of small acid drops, or lemon sherbet, etc.

Proprietary stomach powder or Milk of Magnesia

Lemon juice

Vinegar

Citric acid

(Continued)

Procedure

In this experiment the pupils first test a solution made from an acid drop or lemon sherbet and then one from stomach powder or Milk of Magnesia. From the acid drop, pupils can move on to other acidic materials, such as lemon juice and vinegar. Full-range Indicator can then be introduced as a substance that not only indicates whether a substance is acidic but also how acidic it is.

Now they can take a series of solutions and see what happens when Full-range Indicator is added to each solution. How are we to specify the different acidities of these solutions? It would be clumsy continually to refer to the colour of the indicator. We therefore use a scale of numbers – the pH scale. The pupils may be told that there is a quantitative definition which determines this scale, but for the time being it will be used simply as a way of expressing acidity and alkalinity, alkalinity being at this stage simply the property of any solution which has a pH greater than 7. A solution of pH 7, which is the 'acidity' of pure water, is called a neutral solution.

Calcium hydroxide (slaked lime)

Extract of plant material prepared as in Experiment B2.1a (or pupils' own extract)

Various acidic and alkaline solutions as available, including 0.1M hydrochloric acid and 0.1M sodium hydrogen carbonate

Full range-Indicator (BDH)

The lesson may now be broadened in two ways; other substances may be tested and other coloured materials used as indicators. Among the other substances to be tested are lemon juice, vinegar, lime, and any household substance such as detergents, bleaches, and so on. At this stage, some of the chemicals they have come across in the course, such as alum, may also be tested.

If there is time, you may like to finish up by giving them three or four 'mystery' solutions (for example, dilute solutions of hydrochloric acid, sodium acetate, ammonium chloride, and ammonium hydroxide), asking them to find the pH of each solution with Full-range Indicator.

Experiment sheet 5

In this experiment you will study the effect of a number of chemicals on coloured substances extracted from plants. You may have already prepared solutions of some of these coloured substances. If not use the method of *Experiment* 4 with ethanol (alcohol) as a solvent instead of acetone.

Put six drops of each coloured solution that you have prepared in the middle of separate pieces of filter paper and allow them to dry.

While waiting dissolve a small piece of an acid drop or some lemon sherbet in a little water and add four drops of one of your coloured solutions.
What happens?

Now test a solution of an 'anti-acid' substance, such as stomach powder, with the same coloured solution.
What happens this time?

Test each of the substances named in the table below, as follows: add a measure* of the solid substance or ten drops of

	Extract prepared from		
Colour of extract alone			
Colour with lemon juice			
vinegar			
citric acid solution			
bicarbonate of soda solution			
lime solution			

*A 'measure' means the amount of solid that would lie on a new penny, or would just fill the grooved end of a Nuffield spatula.

liquid to about 5 cm^3 distilled water and shake well. Using a teat pipette, place one drop of the solution obtained near the edge of one of your coloured extracts on the filter paper. Make notes of all colour changes in the table. Repeat with other coloured extracts.

A more sensitive indicator

Better 'indicators' for acids and alkalis are available. One of these is called Full-range Indicator. This changes colour according to the acidity or alkalinity of the solution. Acidity or alkalinity is given a number (called the 'pH value') according to the colour of the indicator. The lower the number the more acidic the solution, the higher the number the more alkaline. A table will help you to understand.

Colour of Full-range Indicator	Acidity/alkalinity number (pH value)	
Red	1	↑
	2	
Orange	3	increasing acidity
	4	
Orange-yellow	5	
Yellow	6	
Greenish-yellow	7	neutral
Yellowish-green	8	
Green	9	
Bluish-green	10	increasing alkalinity
	11	
Blue	12	
	13	
Violet	14	↓

You can use Full-range Indicator paper to test various liquids and solutions for acidity and alkalinity. Put a small drop of the solution to be tested on a piece of indicator paper and compare the colour with the chart provided with the paper. Record your results in the table below.

Substance tested	Colour with indicator	pH value

New words

Acid (adjective, acidic)
Alkali (adjective, alkaline)
Neutral
Indicator

Suggestions for homework

Use some samples of Full-range Indicator paper to test household substances such as milk, detergent, toothpaste, and vinegar. Note the results and bring them back for the next lesson.

Summary

By the end of this section the pupils should have a working knowledge of what is meant by acidity and alkalinity and be able to refer to degrees of acidity–alkalinity in terms of pH numbers. They should know how to use Full-range Indicator to find the pH of solutions, and that solutions having pH numbers less than 7 are called acids; those with pH numbers greater than 7 are called alkalis; and a solution of pH 7 is known as neutral.

B2.2
How can acidity be cured?

The pupils are now familiar with the idea of acidity–alkalinity and have used the pH scale. In this lesson the principle of titration is introduced.

A suggested approach

Objectives for pupils

1. Knowledge of how to neutralize acidity
2. Skill in a simple form of titration
3. Recognition that temperature changes in reactions are worth noting
4. Understanding of the meaning of the word neutralization

Start the lesson by discussing the results of the suggested homework from B2.1 if it was set. From this, lead into a discussion on the soil. Some of the pupils will have heard that gardeners and farmers say that lime can 'cure' acidity in the soil. They have seen that the acidity of acid drops can be eliminated by adding an alkaline substance. Take a sample of soil known to be acid, stir in a little water, filter or centrifuge, and test the filtrate with Full-range Indicator. It will show a pH number in the acidic range. Now add some lime to the soil and repeat. The acidity has gone. How much lime must the farmer add to the soil? If he adds too much the soil will become alkaline. There is an optimum pH for crops and it varies from crop to crop. Obviously we must find some way of following the pH changes as lime is added. Discuss with the pupils how this may be done. An experiment in which lime is added in small quantities to vinegar is described below.

Experiment B2.2

To investigate the effect of an alkali on an acid

Apparatus

Each pupil (or pair) will need:

Experiment sheet 28

Beaker, 100 cm^3

Measuring cylinder, 25 cm^3

Glass rod, or robust −10 to +110 °C thermometer

Spatula

Vinegar

Slaked lime

Small pieces of Full-range Indicator paper

Procedure

In this experiment the pupils use Full-range Indicator paper to follow the change in pH when portions of slaked lime are added to vinegar.

The most convenient way of finding the pH of the solutions is to withdraw a drop on the end of a glass rod and place it on a small piece of Full-range Indicator paper.

If thermometers are used as stirrers, the pupils can follow the temperature changes throughout the reaction (a 4 °C rise is typical).

Note. There may be small variations in the acidic content of different vinegars. It is advisable to check the amount of slaked lime needed before the lesson and adjust the concentration of the vinegar accordingly. About six spatula measures of slaked lime are sufficient for the pupils to have to use; more may make the experiment tedious.

Vinegar substitute. 5 cm^3 of 2M acetic acid requires about seven spatula measures, producing a temperature rise of approximately 4 °C.

Ammonia as an alternative to slaked lime. 10 cm^3 of vinegar requires five 1 cm^3 portions of 4M ammonia solution added by teat pipette, giving a temperature rise of about 4 °C.

Details of the suggested procedure are given in *Experiment sheet* 28 reproduced below.

Experiment sheet 28

Use the Full-range Indicator to find what happens when an alkali is added to an acid.

Put about 10 cm^3 vinegar in a small beaker and add roughly the same volume of distilled water.

You are going to find out the effect on the acidity of the diluted vinegar by adding small portions of lime. How can you tell whether the acidity (measured by the pH value) is changing?

Use your method to measure the pH of the diluted vinegar solution in the beaker. Record this value by putting a large pencil dot on the vertical (pH) axis on the graph below.

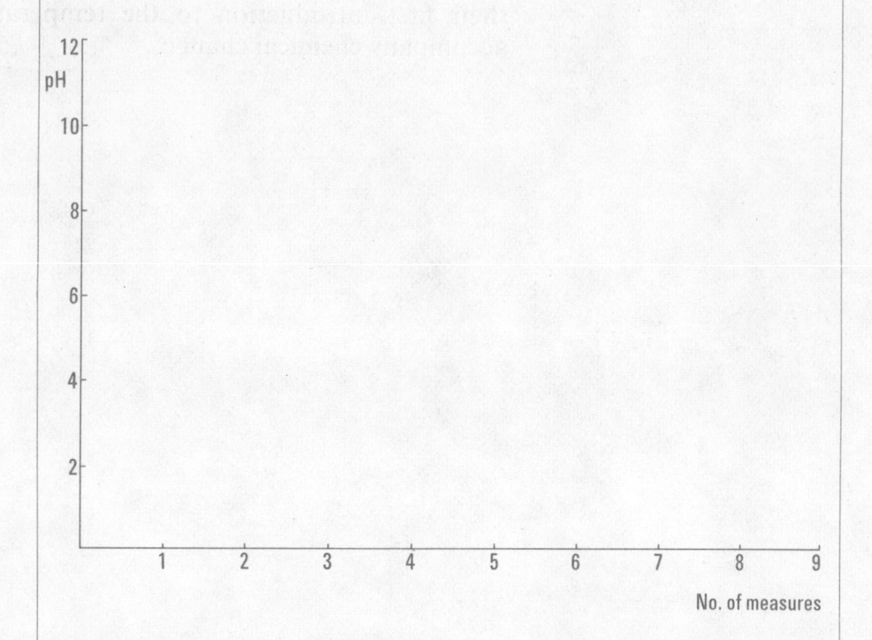

Add one measure of powdered lime to the beaker, stir well, and measure the pH of the mixture. Put a dot vertically above the '1' on the horizontal (number of measures) axis on the graph to represent this pH value. Add another measure of lime, stir again, measure the new pH value and record this above the '2 measures' mark on the graph. Continue in this way until no further change occurs.

Experiment sheet 28 *(continued)*

Join the dots together with a line and look at the shape of the graph obtained. Can you explain it?

Now find out whether the original indicator colour can be got back by adding more vinegar. Record what happens below.

After the experiment discuss the results and draw attention to important details such as the rise in temperature during the reaction.

Introduce the word 'neutralization' to describe the action of the alkali on the acid.

New word

Neutralization

Suggestions for homework

1. Why is the process of 'curing' acidity important in the following cases?
a. Treating acidic soil.
b. Cleaning your teeth.
c. Curing stomach ache.
2. Make a summary of the work you have done in this Topic.

Summary

After this section of work, pupils should know that an alkali will neutralize an acid, and thus be suitable for 'curing' acidity, and should have some skill in a simple form of titration. They have had their first introduction to the temperature changes that often accompany chemical changes.

Topic **B3**

Fractional distillation as a way of separating mixtures

Purposes of the Topic

In this Topic the use of distillation as a method of separating mixtures, already met in Topic B1, is developed further. The purpose is:

1. To show how crude oil is separated into fractions (which can be done experimentally by the pupils using simple apparatus).
2. To show how the components of air may be separated by the fractional distillation of liquid air (this separation being studied on film). This leads to knowledge of the composition of the air.

Contents

B3.1 How can a mixture of two liquids be separated?
B3.2 Fractional distillation of crude oil
B3.3 Fractional distillation of air

Timing

The first two sections will probably need a double period each; the third could be completed in one single period. To allow time for discussion, three double periods should be allowed for the Topic as a whole.

Introduction to the Topic

In Topic B1 the substances separated from one another were solids or solids and liquids. In this Topic the first problem is again one of separation or purification. 'How can we separate two liquids?' The answer lies, as the pupils expect, in distillation, but distillation of a more sophisticated kind. They see ethanol separated from water and then themselves separate some fractions from crude oil. The purpose behind this is to approach the problem of the composition of the air in a novel way. Air can be liquefied and likewise fractionally distilled. It can therefore be shown to be a mixture, mainly of nitrogen and oxygen. The properties of these gases are investigated in the next Topic.

By the end of this Topic the pupils should have a working knowledge of the process of fractional distillation. They will not be expected to understand the theory of fractionation, which is more suitable for the sixth form. They should know that air is a mixture containing mainly nitrogen and oxygen and the fact that it can be liquefied and separated by fractional distillation is evidence for this. The film loop 1–6 'Liquid air fractionation' will have shown them that air consists of four-fifths nitrogen, one-fifth oxygen.

Background knowledge

In Topic B1 pupils will have formed some idea of what is meant by a single substance, and will have met simple distillation as a means of purifying sea water, that is, a means of separating water from dissolved solids.

Subsequent development

Topic B4 on the properties of the major gases of the air leads directly from this Topic. Topic B9 considers burning and breathing.

Supplementary materials

Film loops
1–3 'Whisky distillation'
1–4 'Oil prospecting'
1–5 'Petroleum fractionation'
1–6 'Liquid air fractionation'. A chart on liquid air fractionation accompanies this film loop.

Films
'Refining'
'Oil'
'North Sea strike'
'O for oxygen'
'Oxygen in steel making'
'Air'
See Appendix 4 for brief descriptions and further details

Reading material

for the pupil

Study sheet:
Where chemicals come from describes how the chemicals the pupils use, and many others besides, are obtained from air, sea, rocks, and plants. It covers a variety of substances and its aim is to widen the children's appreciation of the extraordinary wealth of Nature.

B3.1
How can a mixture of two liquids be separated?

The idea of distillation is extended to try to separate two liquids, and fractional distillation is developed. An ethanol–water mixture is used for the experiment.

A suggested approach

Objectives for pupils

1. Awareness of the technique of fractional distillation as a means of separating mixtures of liquids
2. Knowledge of the industrial use of distillation, as exemplified by the production of whisky

The teacher will need

Film loop projector

Film loop 1–3 'Whisky distillation' (with notes)

The idea of purification by separation will be familiar to the class from Topic B1. Start this section by referring to all the ways in which substances were separated from each other in that Topic. One of the methods, distillation, was used to separate a solid from a liquid. Ask the class how they think two liquids might be separated. The obvious answer is distillation, but there are difficulties. What if the two liquids have boiling points which are near to each other? At this point it might be worth introducing the subject of brewing and the distillation of alcohol (ethanol) in industry. To make 'spirits' a fermented brew containing alcohol and water has to be distilled. Explain that a special type of distillation apparatus has to be used, and introduce the fractionating column. As an example of fractional distillation you may fractionate a homemade glucose–yeast brew or simply use a mixture of ethanol and water. Full details of this experiment are given below.

Experiment B3.1

Fractional distillation
This experiment should be done by the teacher.

Apparatus

The teacher will need:

Round-bottom flask, about 500 cm^3

(Continued)

Procedure
Demonstrate fractional distillation by setting up the apparatus as shown in the diagram. The fractionating column may be packed with small pieces of glass rod or glass tubing. Important points to note are:

Fractionating column and packing

Liebig condenser with rubber tubing to connect to water supply and sink

Bunsen burner, asbestos square, tripod, and gauze

Thermometer, -10 to $+110\,°C$ in cork or bung* to fit fractionating column

Several beakers, 100 cm^3, to collect distillates

3 stands and clamps

Corks or bungs to assemble apparatus*

Anti-bumping granules

Mixture of ethanol (industrial methylated spirit) and water, one part of ethanol to three parts of water

1. The round-bottom flask should not be more than a third full at the beginning of the distillation.
2. The thermometer bulb should be just below the side-arm at the top of the fractionating column, to record the temperature of the vapour actually being collected.
3. The cooling water should enter the condenser through the lower tube and leave by the upper one, so that the condenser jacket is always full.

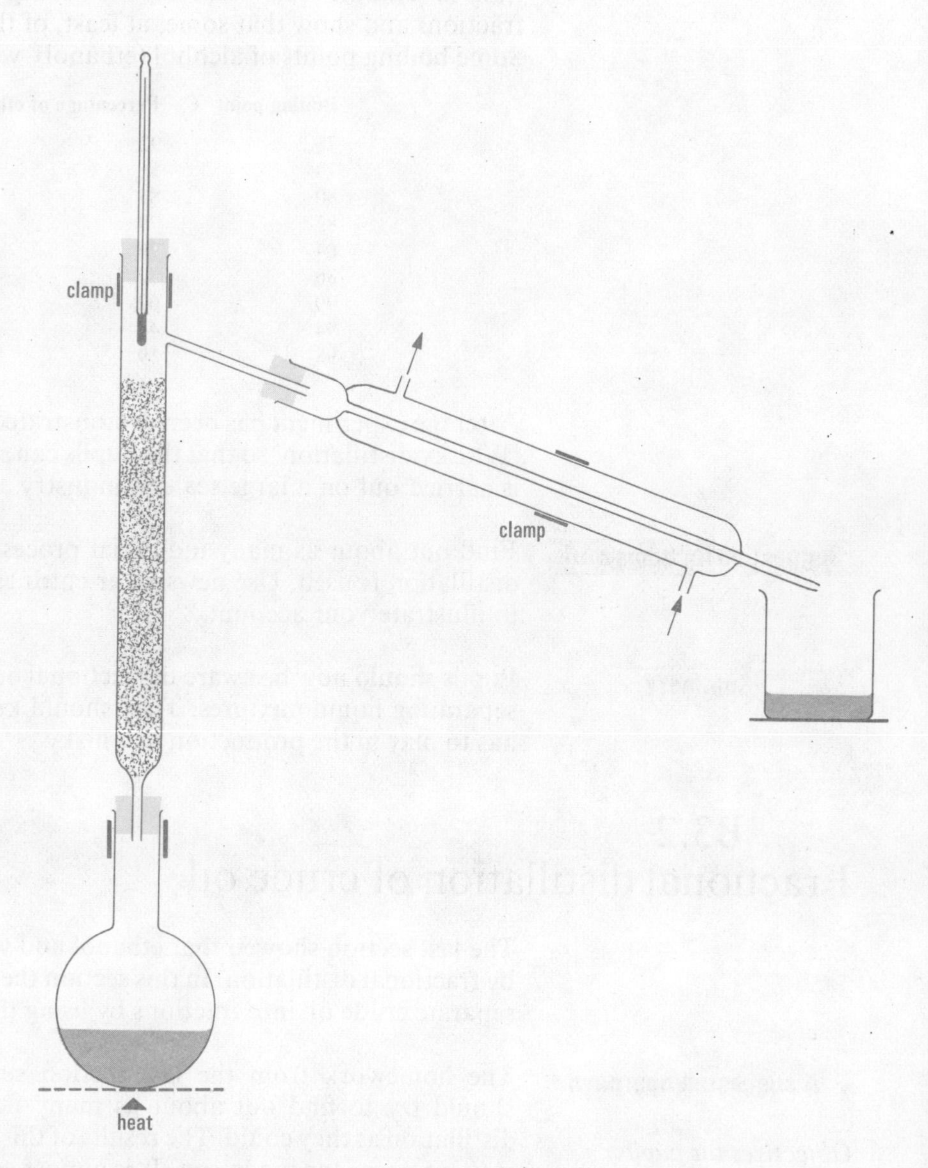

*Corks or bungs will not be needed if apparatus fitted with interchangeable ground-glass joints (for example, Quickfit apparatus) is used. Although more expensive, this type of apparatus is much more convenient, and is recommended for the teacher's own use.

Before starting the experiment, show that the mixture of water and ethanol will not burn.

Place a few anti-bumping granules in the round-bottom flask to promote even boiling. Boil the liquid in the flask so that the distillate collects at a rate of two to three drops a second. Take care to avoid boiling too rapidly otherwise the column may 'flood' (become filled with liquid) and a quantity of liquid may be forced over into the condenser. Watch the thermometer and collect different fractions in separate receivers. Note the boiling ranges of these distillate fractions and show that some, at least, of these fractions will burn. Some boiling points of alcohol (ethanol)–water mixtures are:

Boiling point/°C	Percentage of ethanol
78.2	92
79	86
80	83
82	79
84	76
86	72
90	62
94	44
98	19

After the experiment has been demonstrated, show the film loop 1–3 'Whisky distillation' so that the pupils can see how the same process is carried out on a large scale in industry.

Suggestion for homework

Find out about as many industrial processes as you can in which distillation is used. Use newspaper cuttings and magazine pictures to illustrate your account.

Summary

Pupils should now be aware of fractional distillation as a method of separating liquid mixtures. They should know the part distillation has to play in the production of whisky.

B3.2 Fractional distillation of crude oil

The last section showed that ethanol and water could be separated by fractional distillation. In this section the pupils find that they can separate crude oil into fractions by using the same technique.

A suggested approach

Objectives for pupils

1. Skill in the technique of fractional distillation
2. Knowledge of the industrial process of crude oil fractionation
3. Understanding of the meaning of the words

The homework from the last section suggested that the pupils should try to find out about as many industrial applications of distillation as they could. The results of this homework make a good starting point for the lesson. It is almost certain that someone will have discovered that an essential part of oil refining is fractional distillation.

Show the pupils a sample of crude oil and start a discussion on its importance and uses. Emphasize that although oil has been used in the past primarily as a source of energy, it is now extremely im-

fractional distillation and (crude oil) fraction

The teacher will need

Crude oil sample
(If necessary an artificial crude oil can be made up by mixing petroleum ether, paraffin, lubricating oil, and a heavier oil.)

Film loop projector

Film loop 1–4 'Oil prospecting' (with notes)

Film loop 1–5 'Petroleum fractionation' (with notes)

Useful films

'Refining'

'Oil'

'North Sea strike'

See Appendix 4 for brief descriptions and further details

portant as a source of chemicals. Compounds from crude oil are used for producing a very large number of substances, among them plastics, detergents, anti-freeze (ethylene glycol), and fertilizers. Where and how is crude oil found? This discussion can be usefully supplemented by showing film loop 1–4 'Oil prospecting'. Before the lesson, carefully read the notes supplied with the loop.

Now turn to the subject of how crude oil can be separated into fractions and suggest the experiment described below.

Note. The following organizations publish information on the production of materials from crude oil (as well as drilling, distillation, etc.). This information is *only* available to teachers, and they must apply on their school writing paper.
Information Service Department, Institute of Petroleum, 61 New Cavendish St, London W1M 8AR.
Shell Information Service, Shell International Petroleum Company Ltd, Shell Centre, London SE1 7NA.
BP Educational Service, PO Box 21, Redhill, Surrey.
Educational Services, Public Affairs Department, Esso Petroleum Company Ltd, Esso House, Victoria St, London SW1.

Experiment B3.2

Fractional distillation of crude oil

Apparatus

Each pupil (or pair) will need:

Experiment sheet 6

Test-tube, hard-glass, with side-arm ('filter tube'), 125 × 16 mm

Bent delivery tube and rubber connection tubing

4 test-tubes, 75 × 12 mm, in rack

Thermometer, 0–360 °C and cork; the cork should be fitted ready for use to avoid breakage

Teat pipette

4 watch-glasses, hard-glass

Bunsen burner and asbestos square

Stand and two clamps

Anti-bumping granules

About 3 cm^3 of crude oil

Procedure
This is described in *Experiment sheet* 6 reproduced below. The teacher should however note the following points.
1. Before the experiment it is wise to try it out yourself, as samples of crude oil vary, and it may be desirable to alter the temperature ranges over which the various fractions are collected.
2. During the experiment pupils' apparatus should be checked to see that the bulbs of their thermometers are in line with the side-arms of the test-tubes so as to record the temperature of the vapour actually being collected. A supply of tissues to mop up spilt oil is useful.
3. At the end of the experiment the teacher may prefer to demonstrate the tests for inflammability on the various fractions because of the risk of fire, and the sooty nature of the flames.
4. A strong, hot detergent solution is best for washing up the test-tubes afterwards.

Experiment sheet 6
You have been doing experiments with coloured substances obtained from plants. In this experiment you will examine crude oil (or petroleum), one of the most important sources of both fuels and chemicals, to see what can be got from it by heating.

Put three or four anti-bumping granules in the bottom of a test-tube. Place about 3 cm^3 of crude oil on them by means of

Experiment sheet 6 (continued)

the teat pipette. Be careful to keep the oil off the upper part of the test-tube, as this will make it difficult for you to see what happens later. Clamp the tube at an angle and close it with a cork carrying a thermometer. Warm the bottom of the tube with a 2 cm Bunsen flame.
What do you notice in the upper part of the tube?

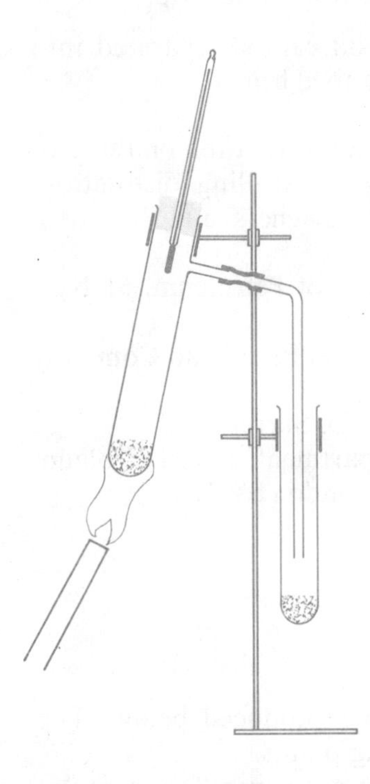

Hold a test-tube to catch any liquid that drips out of the side-arm. When the thermometer reads about 70 °C, collect the liquid in a different test-tube. Again change the test-tube when the temperature has risen to about 120 °C, and again at 170 °C. Collect the last fraction until the temperature is 220 °C. You may need a rather larger flame to drive off the higher-boiling fractions.

When you have finished, you should have collected samples of liquids that boiled over the following ranges:
1. Room temperature to 70 °C
2. 70–120 °C
3. 120–170 °C
4. 170–220 °C

With each of these four test-tubes containing small quantities of materials obtained from the oil, try the following two simple tests:

a. Slowly pour the contents of each test-tube into separate watch-glasses. Compare the ease with which each sample flows. Which flows the most easily?
Which flows least easily?
b. Try to light the liquids in the watch-glasses. You may find it helpful to use a small tuft of cotton wool as a wick. What happens?
Do you think each fraction you have collected is a pure substance?

Pupils could also be told that adding fractions to asbestos paper before placing in a flame gives a good test of flammability.

After the experiment, discuss the results with the class. Tell the pupils that this method of distillation, whereby two or more liquids are separated into fractions (that is, parts of the whole), is called fractional distillation. Have they shown that crude oil is not a single substance?

They will see that they have, but do they think that the fractions they have collected are pure substances? If they were to try to separate them again, using a more sophisticated process, they would find that the fractions contained many different chemicals. Tell them that the process which they have carried out is conducted on a very large scale in industry. Describe briefly how crude oil is fractionally distilled in oil refineries and how, after further processing, the fractions are used for different purposes. Film loop 1–5 'Petroleum fractionation' may be shown at this point. Full details of the film are given in the notes supplied.

New words

Fractional distillation
Fraction (as in crude-oil fraction)

Suggestions for homework

(See *Note* on page 167 about useful booklets.)

1. Read the Study sheet *Where chemicals come from.*
2. Write an account of the uses of fractionation and illustrate your answer with cuttings from magazines and papers.
3. Make a list of things made by processes which involve fractionation.

Summary

After this section the pupils should have some skill in the simple technique of fractional distillation used in Experiment B3.2 (*Experiment sheet* 6) and should know a few simple facts about the industrial methods of separating crude oil, and the uses to which the fractions may be put.

B3.3
Fractional distillation of air

Practical experience with fractional distillation from the last two sections is used to introduce the subject of fractional distillation of liquid air.

A suggested approach

Objectives for pupils

1. Knowledge of the composition of the air
2. Knowledge of the industrial process of liquid air fractionation

The teacher will need

Film loop projector

Film loop 1–6 'Liquid air fractionation' and chart

Useful films

'O for oxygen'
'Oxygen in steel making'
'Air'
See Appendix 4 for brief descriptions and further details

Introduce the subject of the air. Is air a mixture or a single substance? Can fractional distillation help us to find out? Before attempting to separate it, air would have to be liquefied. How can this be done? Suggest cooling the air, and do the following experiment.

Experiment B3.3

Apparatus

The teacher will need:

Gas meter and pump (a filter pump will do)

(Continued)

To find out if anything can be separated from air by cooling it

Suppose we cool down some air using a freezing mixture. We can get a rough idea of how much air we use by including a gas meter in the apparatus. Draw a slow stream of air through a gas meter for about five minutes and then through a U-tube immersed in a freezing mixture. Dry the outside of the U-tube and examine the contents. What does the tube contain? How can we find out whether our suggestion is correct? Allow the product to melt and find the boiling

U-tube surrounded by freezing mixture in beaker

Test-tube, 150 × 25 mm

Thermometer, −10 to +110 °C

point. This leads us to the conclusion that air contains water vapour. It cannot contain very much since we had to use a lot of air to get a little water.

If we were to use much more efficient cooling methods, we would find a small proportion of another solid appearing at about −80 °C. This is solid carbon dioxide, which has a number of uses, one of which is to keep ice-cream cold during transport. It is formed from carbon dioxide gas, which is present in air in much smaller proportions than water vapour.

On further intensive cooling, the air which remains would liquefy at about −200 °C. Using special plant, this liquid can be fractionally distilled. A simplified version of the methods used is shown in the film loop 1–6 'Liquid air fractionation'.

Before showing the film loop it is worth spending some time on describing the phenomena on which the liquefaction of air is based. The process of compression, cooling, and expansion is essentially that employed in the liquefaction of air, the only important addition being that the gases on expanding are made to do work by driving an expansion engine. This cools the gas still further and recovers some of the energy for use again at the compression stage. Pin up the wall chart 'Liquid air fractionation' which is supplied with the film loop and go through the main processes, describing them very simply.

Then show the film loop. Full instructions on how to use it are given in the notes. The film shows that two main products can be obtained, liquid nitrogen and liquid oxygen, and that there is present about four times as much nitrogen as oxygen. These gases can be stored, under pressure, in thick-walled metal cylinders.

Finally, show the class oxygen and nitrogen cylinders, if you have them, calling attention to colour codes and action of valves. These form convenient sources of the gases; they will be used in the next Topic to obtain samples of nitrogen and oxygen for further study.

Suggestion for homework

Make a summary of the work you have done in this Topic.

Summary

Pupils should now know that air consists mainly of nitrogen (4/5) and oxygen (1/5). They should know that these gases can be separated by liquefying the air and fractionally distilling the liquid; and that they can be stored for later use by compressing them inside thick-walled metal cylinders.

Topic B4

The major gases of the air

Purposes of the Topic

To investigate the properties of oxygen and nitrogen in order to lead the pupils to the discovery that substances burn in oxygen but not in nitrogen. The products of combustion of elements are then studied in order to introduce the idea of classifying elements.

Contents

B4.1 What are the properties of oxygen and nitrogen?

Timing

Three double periods are recommended for this Topic. There is a great deal of experimental work to be done and the pupils should do as much of it as possible themselves. Much of the third double period will be needed to bring together the results of the experiments done in the first two periods.

Introduction to the Topic

In this Topic the properties of nitrogen and oxygen as separate gases are investigated. The pupils discover that substances burn much more brightly in oxygen than in air and that, of the substances that are tried, only magnesium burns in nitrogen. They deduce therefore that oxygen is the active constituent of air. The products of combustion with oxygen are also studied. These new substances formed by burning are dissolved in water and the resulting solutions tested with Full-range Indicator. The elements burnt are thereby divided into categories.

Background knowledge

This Topic follows directly from Topic B3, in which the fractional distillation of liquid air was used to show that the gases oxygen and nitrogen are the major constituents of air.

Subsequent development

The use of indicators, and the meaning of the words acidic and alkaline (Topic B2) are assumed during the investigation of the products of combustion of elements.

Further properties of oxygen are examined in Experiment B8.1a, but the main development of the ideas mentioned in this Topic takes place in Topic B9, in which burning, breathing, and rusting are investigated.

Supplementary materials

Films
'Combustion'
'Fire chemistry'
'The air, my enemy'
See Appendix 4 for brief descriptions and further details

Reading material

for the pupil

Study sheet:
Fresh air? deals with some topical problems to do with the air, such as pollution, in a form simple enough to allow pupils to discuss the questions involved and to make them feel at ease in dealing with scientific concepts.

B4.1
What are the properties of oxygen and nitrogen?

In the last Topic it was seen that the air was mainly a mixture of nitrogen and oxygen. The properties of these two gases are now investigated by burning substances in them.

A suggested approach

Objectives for pupils

1. Knowledge of oxygen as the active constituent of the air
2. Skill in performing combustion experiments
3. Awareness of the alkaline or acidic nature of the combustion products of metals and non-metals
4. Knowledge of the limewater test for carbon dioxide

Remind the pupils that all the chemical changes that they have observed so far have been taking place in the atmosphere. Has the air been taking a part in these changes? Certainly many everyday substances change when they are left outside: iron rusts, aluminium tends to go dull and sometimes acquires white powdery marks, and copper-covered roofs or domes go green. To check this a number of substances, say a piece of iron, a piece of aluminium, a piece of copper, some polythene, and so on, may be left set up (1) in the laboratory, and (2) outside, to see what effect the atmosphere has on them.

To investigate more specifically the properties of oxygen, the pupils should be allowed to try to burn different substances in test-tubes full of the gas. If the substance does burn in the gas, the product should be tested in each case with Full-range Indicator.

Details are given as Experiment B4.1 (*Experiment sheet* 12) below. When this experiment is over the teacher can then attempt to burn the same substances in nitrogen.

Experiment B4.1

Burning substances in oxygen

Apparatus

Each pupil (or pair) will need:

Experiment sheet 12

4 test-tubes, 150 × 25 mm, and corks

Test-tube rack

Tongs

Combustion spoon

Bunsen burner and asbestos square

Teat pipette

Stick or small piece of wood charcoal

Steel wool

Magnesium ribbon (about 3 cm)

Full-range Indicator solution

Powdered sulphur

Limewater

(Continued)

Procedure

The first thing to be done is to distribute samples of the gas to the pupils. This may be done in one of two ways:

1. The teacher uses a trough or plastic washing-up bowl three-quarters full of water. The pupils bring their test-tubes to the bowl, fill them with water and turn them upside down, with the open end beneath the surface of the water, over the end of a rubber delivery tube which leads the gas from the cylinder to the bowl. The gas displaces the water in the test-tube and as soon as it is full the pupil places a cork in it and returns to his bench. It saves time if a number of test-tubes are filled before the lesson begins.
2. Small polythene bags, closed by a bung and delivery tube, may be used over short periods of time for storing gases. They are inflated from a gas cylinder just before the lesson and are suitable for class use when the pupils need several test-tubes of oxygen.

If this second method is used the following will be needed: polythene bag, approximately 25 × 15 cm; rubber bung, approximately 2.5 cm diameter, fitted with 5 cm glass tubing carrying approximately 5 cm of connection tubing; elastic band; and Mohr's clip.

A groove is cut out round the bung, sufficiently deep to accommodate the elastic band. The mouth of the polythene bag is fastened round the bung and secured to it by the elastic band. All the air is

The teacher will need:

Length of rubber tubing to lead oxygen from cylinder to pupils' apparatus

Cylinders of oxygen and nitrogen

Calcium turnings

For an alternative means of gas collection each pupil should have two polythene bags and a small trough.

expelled from the bag, before closing it with the Mohr's clip. The empty bag is lightly connected to the tubing from the cylinder and the Mohr's clip held open. When the bag is almost inflated, stop the gas flow by the fine adjustment valve, close the Mohr's clip and disconnect the bag.

The gas may be collected over water in 150×25 mm test-tubes by fitting a delivery tube and gently expelling the gas from the bag. This size of bag would give about eight 150×25 mm test-tubes of gas.

Note. If an oxygen cylinder is not available, or the teacher prefers to let each pupil prepare his own supply of oxygen, a convenient method is by the addition of 20-volume hydrogen peroxide solution to granules of manganese dioxide, using the apparatus shown in Experiment A4.1, page 82.

When the pupils have collected some test-tubes of oxygen they can then investigate the burning of the elements mentioned in *Experiment sheet* 12, which is reproduced on the next page.

Experiment sheet 12

Many oxides can be made by burning substances in oxygen. In this experiment four oxides are made in this way and some of their properties studied.

Collect four large test-tubes full of oxygen. When you have filled each tube, cork it securely, and put it in a test-tube rack.

The substances to be burnt can be held in a small pair of tongs (apart from sulphur) or placed on a small combustion spoon. Enter your observations in the table below.

1. *Carbon*. Heat to redness one corner of a piece of charcoal. Remove the cork from one test-tube of oxygen and at once insert the hot charcoal. When the action is finished remove the charcoal and replace the cork in the tube.
Perform the following two tests on the gas now in the tube.
a. Remove a sample of the gas with a teat pipette. Expel the gas through a little water containing two to three drops of Full-range Indicator in a small test-tube. Record the colour of the indicator.
b. Remove another sample of the gas into the teat pipette and bubble it through a little clear limewater in another small test-tube. Record any change in the limewater. (This is an important test; you will often need to use it later.)

2. *Iron*. Use a small piece of steel wool or some iron powder. Heat it until it is red hot and plunge it into a test-tube of oxygen. Shake the product of burning with a little water and add a few drops of Full-range Indicator solution.

3. *Magnesium*. Use about 5 cm of magnesium ribbon held in tongs or wound round a combustion spoon. Set one end of the ribbon alight and plunge the burning metal into oxygen.
Do not look directly at the burning magnesium, it can cause damage to the eyes. Shake the product with a little water and add Full-range Indicator solution.

4. *Sulphur*. Light the sulphur first, in a combustion spoon or picked up on a moistened pipe cleaner or asbestos paper strip. Plunge it into oxygen. Test product with Full-range Indicator and limewater separately as for the product from carbon in (1).

Substances burnt	Observation during burning	Name of product	Appearance of product	Indicator colour	Action of limewater
Carbon					
Iron					
Magnesium					
Sulphur					

When the pupils have completed their part of the experiment, the teacher should show them what happens when calcium is burnt in oxygen. Hold a calcium turning in a small pair of tongs and carefully scrape off the surface layer of oxide. Still holding it in the tongs, ignite the turning in a roaring Bunsen burner flame, and hold it in a test-tube of oxygen.

The resulting calcium oxide should be dissolved in water and the resulting liquid filtered to give a clear solution (limewater, calcium hydroxide solution). This should then be poured into a test-tube in which carbon has been burnt (yielding carbon dioxide) when a white precipitate is formed. Pupils should be told that the use of limewater in this way is a test for carbon dioxide.

Finally the teacher should attempt to burn all the substances used in the experiment in nitrogen.

After the experiment has been completed, gather the results together and summarize them on the blackboard. The pupils will have discovered from their experiments that most of the substances they were given burned in oxygen but not in nitrogen. When they dissolved the oxides in water they found that solutions had different pH numbers. This information could be correlated by drawing a pH chart on the blackboard from 0 to 14 and marking the positions of the oxides of the various elements on it. From this they gather that the metals come on one side and that the non-metals come on the other. The theory of elements, and of metals and non-metals, should not be discussed at this stage, as the subject is dealt with in Topic B9.

The energy transfers accompanying the preparation of oxides are worth mentioning here, perhaps on the lines indicated in Chapters 15 and 20 of the *Handbook for teachers*.

Suggestions for homework

1. Read the Study sheet *Fresh air?*
2. Find out about as many uses as you can for oxygen, nitrogen, and the other gases in the air.

Summary

By the end of this Topic the pupils should know that oxygen is the active constituent of the air, that oxides can be divided into categories according to the pH of their solutions in water, and that heat is evolved on burning. They will also know how to handle gases in the laboratory.

Topic B5

Finding out more about substances by heating them

Purposes of the Topic

To investigate the action of heat on a number of substances to find out something of their chemical nature.

Contents

B5.1 What happens when substances are heated?
B5.2 A further look at the effect of heating copper sulphate crystals

Timing

Two double periods may be needed for the first section and one for the second. Allowing time for discussion and for following up pupils' suggestions, four double periods should be satisfactory. If time is short, cut down the number of substances to be heated in B5.1 and condense the whole Topic into two double periods.

Introduction to the Topic

In this Topic the effect of heating substances is investigated by finding out whether and how they change in weight*. This leads to the particular study of the effect of heat on hydrated copper sulphate. The pupils identify water as one product and note the fact that heat is evolved when the water is added again to the anhydrous copper sulphate.

By the end of the Topic they should understand that there is a difference between heating and burning. They will know that water is evolved when hydrated copper sulphate is heated and will understand how to identify water by adding it to anhydrous copper sulphate. By now they will realize that heat can cause chemical change and conversely that chemical change can produce heat.

Alternative approach

A more elaborate investigation of the action of heat on substances, in which more than one result is followed up in detail, is given in Alternative A, Topic A2.

Subsequent development

This Topic leads on to Topic B7 on elements.

Further references

for the teacher

Further experiments suitable for Topic B5 are to be found in *Collected experiments*, Chapter 1 'The effects of heating substances'.

Reading material

for the pupil

Study sheet:
Heating things shows that when pupils use heat from a Bunsen burner they are doing what, on a much larger scale, is done in industry every day to make such familiar materials as glass, china, and steel. Cooking is also given as an example of how heat can change a substance into something more useful.

*The term *weight* is used in Stage I of this course, but some teachers may prefer to use the term *mass* (as in Nuffield Combined Science), especially if the pupils are following a Physics course in which *mass* is used.

B5.1
What happens when substances are heated?

In Topic B4 a number of substances were heated in oxygen and nitrogen. Most, but not all, of the substances chosen burned in oxygen. In this section, substances are heated in air, and the field is widened to include materials that behave in other ways. Some of the substances chosen do in fact burn; others decompose, and yet others remain unchanged. Each material is weighed before and after heating to find out what is happening.

A suggested approach

Objectives for pupils

1. Awareness of the idea of a chemical change
2. Skill in heating substances in test-tubes, and in weighing

Start with a discussion of the results of the last Topic and suggest that it would be interesting to see what happens when other substances are heated in air. In this case the substances chosen do not all burn like the substances in Topic B4. The object is to find out what happens. One way of treating this problem is to let the pupils heat a little of each substance in a hard-glass test-tube first. To save time, each pupil or pair of pupils may be given a different substance to heat. Alternatively, each pupil could be given three or four substances including hydrated copper sulphate. Ask them what they can see happening. They will notice the changes in colour and that in some cases a gas or a liquid is driven off. When they have done this, ask them what they have seen and how they can be sure of what has happened. 'If a gas or liquid came off, how would the weight of the substance change?' This question may lead to the idea that weighing the substances before and after the experiment will tell us more about them. Now let the pupils carry out the experiment described below.

Experiment B5.1

Finding out what happens when various substances are heated

Apparatus

Each pupil (or pair) will need:

Experiment sheet 29

4 hard-glass test-tubes, 75 × 12 mm

Test-tube holder

Spatula

Bunsen burner and asbestos square

Access to laboratory balance capable of detecting changes of 0.005 g

Substances for examination:

Copper foil (or reduced wire-form copper(II) oxide)

Lead(II) oxide (litharge)

(Continued)

Procedure

This is described in *Experiment sheet* 29 reproduced below.

Experiment sheet 29

You have already heated a certain number of substances. Now you are going to find out more about what happens when a range of substances is heated in air, using a balance to help you in your observations.

There is quite a lot to record in this experiment. Make a table on a separate sheet of paper which can be fixed into your folder. It may be more convenient to use the longest side of the paper for this. Use these headings:

Name of substance	Appearance			Weight		Gain (G) Loss (L) No change (NC)
	before heating	during heating	after cooling	before heating	after heating	

Copper(II) sulphate, hydrated crystals

Cobalt chloride, hydrated crystals

Sodium chloride

Copper(II) chloride

Copper carbonate

Lead nitrate

Iron(II) sulphate

For powdered solids use about 1 cm depth in a small test-tube, and weigh tube + substance. Heat with a non-luminous flame about 5 cm high, holding the tube in a suitable holder, sloping slightly upwards from the horizontal. Heat until no further change can be seen. Then allow the tube to cool on an asbestos square and weigh again. If copper foil is used, a piece about 5 cm × 2.5 cm can be held in tongs and heated directly. Record all observations and results in the table.

Discuss the results with the class. Ask them which substances lost weight, which gained weight, and which were unaltered. What happens when copper sulphate crystals are heated? The pupils will have noticed that there is a colour change and some may have seen drops of a liquid condensing on the walls of the test-tube. What is happening? This will be followed up in the next section.

Suggestions for homework

1. Read the Study sheet *Heating things*.
2. Design an apparatus to collect the gas given off when a substance is heated. How would you modify the apparatus to enable a liquid, whose vapour is given off, to be condensed and collected?

Summary

Pupils should now have some skill in heating substances, and in observing the changes that take place; and an idea of the value of weighing. It is not intended that they should *remember* all the results.

B5.2
A further look at the effect of heating copper sulphate crystals

In this section the change that takes place when copper sulphate is heated is studied more carefully. (See B5.1.) In particular the following questions are asked: 'What is the liquid which comes off?', 'What evidence have we that it is water?', and 'What happens when the water and white copper sulphate are mixed together again?'

A suggested approach

Objectives for pupils

1. Awareness that assumptions must be tested by experiment
2. Awareness that energy changes accompany chemical changes
3. Understanding of the meaning of the words anhydrous and decompose

Remind the pupils of the experiment in which they heated copper sulphate crystals (Experiment B5.1). They should have noticed that when blue copper sulphate is heated a colourless liquid condenses on the walls of the test-tube. The first question is 'What is the colourless liquid?' Most pupils will assume that it is water. But how do they know and how can they show that it is in fact water? One way is to measure its boiling point. Many children of this age still have difficulty in believing that there are colourless liquids which are not, or do not contain, water. It is worth demonstrating a simple measurement of the boiling point of a liquid other than water at this stage. Details of such an experiment are given opposite.

Experiment B5.2a

How to measure boiling points

This experiment should be done by the teacher.

Apparatus

The teacher will need:

Beaker, 250 cm^3

Bunsen burner and asbestos square

Tripod and gauze

Hard-glass test-tube, 150 × 25 mm

Thermometer, −10 to +110 °C

Stand and 2 clamps

Anti-bumping granules

Ethanol or acetone

Procedure

Place a 250 cm^3 beaker half full of water on a tripod and gauze, over a Bunsen burner, and by means of a stand and two clamps support a 150 × 25 mm test-tube containing not more than 10 cm^3 of ethanol in the beaker, and a thermometer in the test-tube so that its bulb is held just above the level of the ethanol. Put one or two anti-bumping granules in the ethanol. Heat the water in the beaker to about 80 °C and when the ethanol begins to boil note the steady reading on the thermometer. *Ethanol should not be heated directly by the Bunsen burner because of the risk of fire.* As an alternative, acetone (which is also inflammable) may be used in place of ethanol. Boiling points are: ethanol 78 °C, acetone 56 °C.

If the boiling point of the liquid which we suspect is water is to be measured, an apparatus must be devised to collect the liquid. Ask for suggestions from the class and build up an apparatus similar to the one suggested below. The pupils may now try the experiment themselves. Details are given below.

Experiment B5.2b

Investigating the effect of heat on copper sulphate crystals in more detail

Apparatus

Each pupil (or pair) will need:

Experiment sheet 9

2 hard-glass test-tubes, 100 × 16 mm

Cork or bung to fit test-tube, carrying a delivery tube

Bunsen burner and asbestos square

Thermometer, −10 to +110 °C

Stand and 2 clamps

Beaker, 100 cm^3

Copper sulphate, small crystals

Procedure

When the instructions given in *Experiment sheet* 9 have been carried out, let the pupils see the effect of adding several liquids to anhydrous copper sulphate (examples might be hexane, tetrachloromethane, acetone, and water) and thus introduce the use of anhydrous copper sulphate as another test for water.

Experiment sheet 9

Having observed what happens when a number of substances are heated, you can now follow up one of these changes more carefully.
What happened when you heated a few crystals of copper sulphate in a previous lesson?
In this experiment you are going to try to find out more about this change.

Half fill a test-tube with dry blue crystals and arrange the apparatus as shown in the diagram [overleaf]. Gently heat the crystals with a medium Bunsen flame. What do you see happening to the crystals?
Carry on with the heating until you think the reaction is complete. What do you think you have in the righthand test-tube?
How could you test your suggestions?
Remove the righthand test-tube and clamp it almost vertically. Heat the contents with a small Bunsen flame. Hold

Experiment sheet 9
(continued)

a thermometer in the test-tube so that the bulb is near the top and keep it there for a minute or so. What temperature does it register? . . . °C

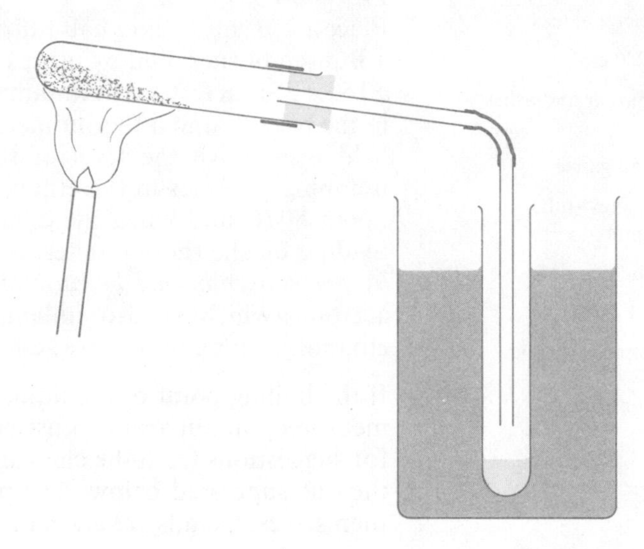

What information does this give you about the contents of the tube?
Do you think the crystals have gone for good? How might you try to re-make them?
Try to do so, using only a small amount of the residue left in the test-tube. What *two* things do you notice happening to the residue?

After the experiment the results should be discussed with the class. Are they satisfied that the liquid which came off was water? Draw attention to the energy changes which took place. They had to put heat energy in to split the copper sulphate into white copper sulphate and water. When the two substances were mixed together again they got some of the heat back ('It got hot'). Explain that the white copper sulphate is called anhydrous ('without water') copper sulphate. Crystalline substances containing water are said to be hydrated.

An informal treatment of reversibility now becomes possible, reinforced by previous experience with indicators. See the *Handbook for teachers*, Chapter 20, page 307.

New words

Anhydrous
Decompose

Suggestions for homework

1. Some substances combine with the air when heated in an open tube, other substances decompose, and some are unchanged, though they melt or evaporate. Look back to the results in your

The pupils in this class were working in pairs, but a small group have collected round the teacher who is discussing with them the action of heat on copper sulphate (Experiment B5.2b).

notes on this Topic and try to decide what happened to each of the materials you heated. As far as you can, explain the reasons for your answers.
2. Make a summary of what you have learnt in this Topic.

Summary

Pupils should now know the meaning of the words decompose and anhydrous (as in anhydrous copper sulphate) and should be able to remember two tests for water, namely (1) determination of its boiling point, and (2) its effect on anhydrous copper sulphate. They should be aware that assumptions must be tested by experiment, and that energy changes accompany chemical changes. (See the *Handbook for teachers*, Chapter 15.)

Topic **B6**

Using electricity to decompose substances

Purposes of the Topic

To show how electricity can be used to decompose materials – as the Bunsen burner was used in Topic B5 – and thus to pave the way to the idea of elements developed in Topic B7.

It is not intended that any theoretical explanation for the observations should be attempted at this stage. A possible approach is outlined in Chapter 6 of the *Handbook for teachers*.

The experience gained in this Topic will be the basis of developing important concepts later in the course, but at this stage the emphasis is on the general aims that pupils should (1) 'Acquire basic knowledge about the behaviour of substances', and (4) 'Develop manipulative skill in laboratory procedures with common apparatus'.

Contents

B6.1 Investigation into substances which conduct electricity
B6.2 Further investigation into substances which conduct electricity
B6.3 What happens when solutions conduct electricity?
B6.4 Using electricity for plating
B6.5 Electricity from chemical reactions

Timing

Sections B6.1 and B6.3 will need two double periods each and the rest will need one double period each. A maximum of eight double periods should be allowed to give time for discussion. If time is short, concentrate on B6.1, B6.3, and B6.4. This reduces the time needed for this Topic to about four double periods.

Introduction to the Topic

This Topic contains a broad examination of the effect of electricity on chemicals. In it electrolysis, electroplating, and the use of chemical reactions to produce electricity are all investigated. In this wide field the depth to which the pupils go is strictly limited. They are meant only to investigate the phenomena, record what they see and, with the help of the teacher, correlate the results.

By the end of this Topic they should:
1. Know how to find out whether substances conduct electricity or not, and have some knowledge of the sort of substances which do so.
2. Know that certain substances do not conduct electricity in the solid state but do so in the molten state, and that when they do conduct they decompose at the electrodes.
3. Have observed in the above cases, and in a few cases in solution, that metals tend to be released at the cathode and non-metals at the anode.

No theoretical explanation of electrolysis is given until Stage II.

Alternative approach

An alternative treatment of this subject can be found in Alternative A, where the use of electricity in getting new materials is met later (Topic A8) than in this scheme. Here, electrolysis precedes the idea of elements and is used to help to define them. In Alternative A, the idea of elements comes before electrolysis, and can be used to help the understanding of the latter, to the extent that electricity can be seen to split up compounds. There is no treatment of simple cells in Alternative A.

Background knowledge

Pupils have not previously met the effect of electricity on materials in Alternative B, but they will understand the meaning of decomposition (by heat) from their work in Topic B5.

Subsequent development

The decomposition of materials by electricity (together with thermal decomposition) leads directly to the idea of elements in Topic B7.

The subject is taken further in Stage II.

Further references
for the teacher

Additional experiments suitable for this Topic are given in *Collected experiments*, Chapter 5 'The effect of electricity on substances'.

Reading material
for the pupil

Study sheet:
Chemistry and electricity helps pupils realize that the batteries they use in torches and transistor radios originate from the observations and experiments made by two Italians over 170 years ago. The Study sheet then deals with the uses of electrolysis and with other ways to make electric current, which have led to our enormous consumption of electricity today.

B6.1
Investigation into substances which conduct electricity

Using a simple circuit the pupils find out that, at the voltages used, only certain substances conduct electricity. They examine air and some liquids and solids.

A suggested approach

Objectives for pupils

1. Familiarity with an electrolysis technique for liquids
2. Knowledge of the behaviour of substances in electrolysis
3. Understanding of the meaning of the words electrode and electrolyte

Before introducing the subject of this investigation, say something about electric circuits. (These and electric currents are developed thoroughly in Year 2 of the Nuffield Physics course.) It will only be necessary to say at this stage that the battery is to be treated as a source of electricity just as a Bunsen burner is a source of heat. The flow of an electric current will be detected where practicable by means of a 6 V bulb. Demonstrate the use of the electrode assembly, battery, and bulb and then tell the pupils to use it to find out whether a number of substances conduct electricity. They can see very easily that air, under these conditions, does not conduct electricity. 'But what about all these liquids and solids?' Have the suggested liquids and solids on the demonstration bench and let the pupils collect samples and test them with their apparatus. It may be necessary to get different pairs of pupils to test different substances and to correlate the results after the experiment. On the whole, the more they can test the better.

If each pupil is to try all the substances suggested a considerable degree of organization is necessary. One good way of organizing the class is to set out a number of 'stations' at which perhaps two or three substances for testing are to be found, and have the pupils go from one to another, until they have tested all the samples available. This method has the advantage of requiring much smaller amounts of materials. A d.c. supply and appropriate electrodes should be at each station.

Experiment B6.1

Investigation into substances which conduct electricity

Apparatus

Each pupil (or pair) will need:

Experiment sheet 30

6 V battery or alternative d.c. supply

6 V bulb in holder

Carbon or steel electrodes in holder

3 lengths of connecting wire fitted with crocodile clips

Several small containers, e.g. beakers, 100 cm^3, or hard-glass test-tubes, 150 × 25 mm

Samples of the chemicals given below:

Distilled water

Ethanol

Limewater

Approximately M or 0.5M solutions of:

Sulphuric acid

Copper(II) sulphate

Zinc sulphate

Sugar

Iron, lead, naphthalene, sulphur, polythene, lead(II) chloride, potassium bromide, sodium iodide.
(A selection of any four of these substances – one metal, one salt, sulphur, and polythene – is enough.)

Procedure

Set up the electrical circuit. The source of electricity can be dry cells giving 6 volts, or if a laboratory low-voltage supply is available, this should be arranged so as to give 6 volts. The 6 V bulb mounted in a bulb holder is included to indicate when the current is flowing. Electrodes may be of carbon or steel, mounted in a wooden support so as to keep them at a constant distance apart.

With this apparatus the pupils can test the substances given above to see if they conduct. The beakers or test-tubes should be almost filled with the liquids for investigation. The electrodes must be washed after each experiment.

A faint glow will be obtained with limewater, but none with ethanol, distilled water, or sugar solution. The metals will be found to conduct, but not the salts, polythene, or sulphur.

The procedure is described in *Experiment sheet* 30 reproduced below.

Experiment sheet 30

You have studied already the effects of heat on a number of substances, and investigated the reactions which occurred. You are now going to look at the effects of passing an electric current through a number of substances.

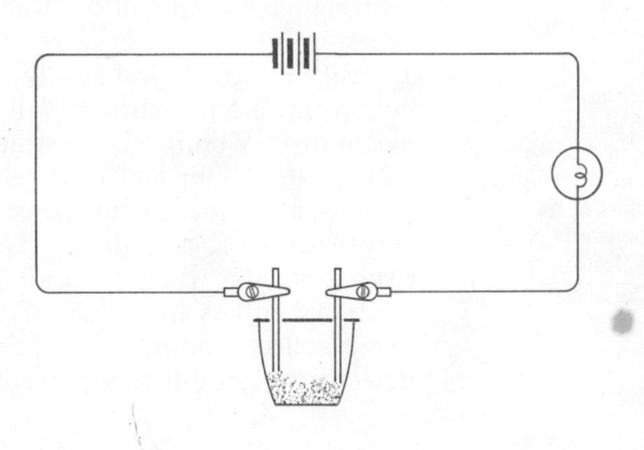

You are going to test various chemicals to see whether or not they conduct electricity. You will already have discussed the pieces of equipment you are going to use and these are not shown in detail. The diagram is only a scheme to show how the different items are to be connected.

The way to find out whether the solid substances conduct electricity is to hold the two electrodes firmly against each substance in turn and see if the lamp lights up.

A small beaker is used to hold the liquids being tested. Try distilled water first, then tap water and the other solutions provided. Make sure that a clean beaker is used each time and that the electrodes are washed before a new solution is used.

Enter the names of the substances in the table below as you test them, and put a tick in the appropriate column.

Substance	Conductor	Non-conductor

The results should be discussed with the class. The substances which conduct fall into two categories: those which conduct and change chemically when they do so, called electrolytes, and those which conduct but do not change. The pupils will have noted that the former type are all liquids, but that not all liquids conduct electricity. 'What happens if solids which do not conduct are melted?' This question will be answered in the next section.

New words

Electrode
Electrolyte

Suggestion for homework

Electricity is a widely used source of power in modern life. Write a description of what you think life would be like without electricity.

Summary

Pupils have now seen how to test both solid and dissolved substances for electrical conductivity. They should know that there exist conductors and non-conductors; and that the former can be divided into electrolytes and non-electrolytes. All electrolytes that they have met are liquid, but not all liquids are electrolytes.

B6.2
Further investigation into substances which conduct electricity

In this section some of the substances which were found not to conduct electricity in Experiment B6.1 are examined further. Certain of them are found to conduct electricity when they are molten.

A suggested approach

Objectives for pupils

1. Familiarity with an electrolysis technique for molten materials
2. Knowledge of the behaviour of certain substances on electrolysis
3. Understanding of the meaning of the words anode and cathode

In the last section of work the pupils found that all the substances which conducted *with decomposition* were liquids. A question which follows naturally from this is: 'Do the solids which do not conduct electricity conduct when they are in the molten state?' Present the pupils with this question and ask them how they would find the answer. Similar apparatus to that used in Experiment B6.1 (*Experiment sheet* 30) will be appropriate, but the pupils will have to think of some way of melting the solid; one answer is to do the experiment in a 150 × 25 mm hard-glass test-tube. As molten salts are involved, and their melting points are high, this experiment should be demonstrated. Pupils can record the results in an extension of the table that they have drawn up for Experiment B6.1. Draw attention to the products at the electrodes (see *Procedure*) and name the electrodes: anode (+ve) and cathode (−ve). A suitable technique is described below.

Experiment B6.2

Finding out if molten substances conduct electricity

This experiment should be done by the teacher.

Apparatus

The teacher will need:

Hard-glass test-tubes, 150 × 25 mm

Stand and clamp

Bunsen burner and asbestos square

2 carbon rods, 20 cm long, and electrode holder

6 V battery or alternative d.c. supply

6 V bulb and holder

3 lengths of connecting wire fitted with crocodile clips

Lead(II) chloride

Potassium bromide

Potassium iodide

Sulphur

Polythene

Procedure

Warning. *Care is needed in handling molten electrolytes.*

Take a sample of one of the substances and fill the 150 × 25 mm hard-glass test-tube to a depth of about 3 cm with it. Now place the electrode assembly in the tube as shown in the diagram so that the ends of the carbon rods dip well into the contents. Connect the circuit, clamp the test-tube in a vertical position, and heat the contents with a Bunsen burner until they are molten.

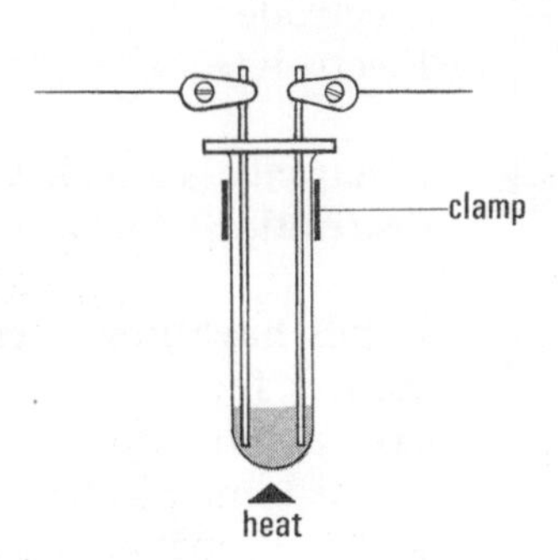

Watch to see if the bulb lights up; it will do so in the cases of lead chloride, potassium bromide, and sodium iodide. Since the bulb limits the current flowing through the electrolyte to about 0.2 amp, it is advisable now to disconnect the bulb from the circuit so as to obtain a larger current. This gives enough action at the electrodes within a few minutes to enable the products to be identified.

The products which can be detected are as follows:

1. *Lead chloride.* A bead of lead can be seen near the cathode and chlorine bubbles off the anode. The pupils should recognize a 'smell

of swimming baths' even if the greenish colour of chlorine is not obvious.

2. *Potassium bromide*. The potassium formed at the cathode is not visible, but at the anode a brown gas with an unpleasant smell (bromine) is evolved.

3. *Potassium iodide*. As with the salt above, the potassium produced is not visible. At the anode a purple vapour (iodine) is evolved.

The electrodes must be clean for this experiment. They easily become contaminated and can then give misleading results. It is suggested that the test-tube and electrodes for a particular electrolyte are always kept together.

Discuss the results with the pupils. They have seen that certain substances conduct electricity in the molten state; these must be added to the list of electrolytes. When the current was flowing there was a certain similarity in the behaviour of these substances. Lead, potassium, and sodium are released at the cathode and chlorine, bromine, and iodine at the anode. Is this pattern followed when solutions of the solids are used? We will try to find the answer to this problem in the next section.

New words

Anode
Cathode

Suggestions for homework

1. Read the Study sheet *Chemistry and electricity*.
2. Describe the way of producing electric current discovered by Michael Faraday.
3. With what part of Volta's work does the following statement of Galvani link up: '. . . the contractions were more violent with some metals than with others'.

Summary

Pupils have now seen how to test molten substances for electrical conductivity. They have seen that some solid non-conductors become conductors on melting and that such substances are all electrolytes. They should know the words anode and cathode, and be aware that in electrolysis metals are formed at the cathode, and other materials at the anode.

B6.3
What happens when solutions conduct electricity?

The investigation is now directed towards solutions to find out what part water plays in electrolysis. The solutions selected are those of materials tested in B6.2 and found to conduct in the molten state (all of which were salts) or similar compounds which are more soluble. The information gained from these two sections is then correlated.

A suggested approach

Objectives for pupils

1. Familiarity with an electrolysis technique for aqueous solutions
2. Knowledge of the behaviour of some solutions on electrolysis

From the last two sections the pupils have learnt two facts: that some solutions conduct electricity and that some substances, normally solid at room temperature, conduct electricity only when they are molten. It looks as if water plays some part in the electrolysis of solutions. Does it do this merely by putting the solid into a 'liquid' phase or is more than this involved? The experiments suggested below are designed to give a simple qualitative answer to this question. Discuss the above points with the class and let them try the experiments. To save time, each pair may try one or two solutions only. Note that the salts given in the experiment are related to those electrolysed in B6.2. Copper chloride and zinc bromide are chosen because they are soluble (unlike lead chloride) and because the copper and zinc released during electrolysis can be clearly seen on the cathode. Before starting the experiment let the pupils see that the salts that they are about to electrolyse in solution really are soluble. They may dissolve small samples in water in a test-tube.

Experiment B6.3

To investigate what happens when various solutions conduct electricity

Apparatus

Each pupil (or pair) will need:

Experiment sheet 31
(Continued)

Procedure

The pupils electrolyse the solutions in the apparatus shown in the diagram in *Experiment sheet* 31 reproduced below. To emphasize the fact that both a solid and a liquid are involved the pupils may

6 V battery or alternative d.c. supply

6 V bulb and holder

Electrolysis cell as shown in the diagram. (For details of its construction, see *Collected experiments*, Experiment E5.3)

3 lengths of connecting wire fitted with crocodile clips

Stand and clamp

Paper tissues or a cloth

Approximately M solutions of:

Copper(II) chloride

Zinc bromide

Potassium bromide

Sodium iodide

Alternatively enough of these solids to allow the pupils to make up their own solutions

Distilled water

make up their own solutions. Once they have shown that the solution conducts electricity, the bulb should be removed from the circuit to allow a larger current to flow. When a gas is given off, collect it in a 75 × 10 mm test-tube and test its properties. The following results will be found using carbon electrodes:

1. *Copper*(II) *chloride solution.* Copper is deposited on the carbon cathode and chlorine is evolved at the anode. The chlorine may be collected and its colour and smell noted, but take care! See Experiment B6.2.
2. *Zinc bromide solution.* Zinc is deposited on the carbon cathode and bromine evolved at the anode. Most of the bromine dissolves in the water, colouring it brown.
3. *Potassium bromide solution.* Hydrogen is evolved at the cathode. Collect it and note that it burns. Note the difference between this and the result in Experiment B6.2. The bromine evolved at the anode dissolves in the water, turning it brown.
4. *Sodium iodide solution.* Hydrogen is evolved at the cathode as above. The iodine evolved at the anode does not vaporize but stains the solution dark brown.

Experiment sheet 31

In the previous experiments you will have noticed changes taking place at the electrodes when solutions are electrolysed. Some of these changes are investigated further here.

Use the apparatus shown in the diagram, with carbon electrodes.

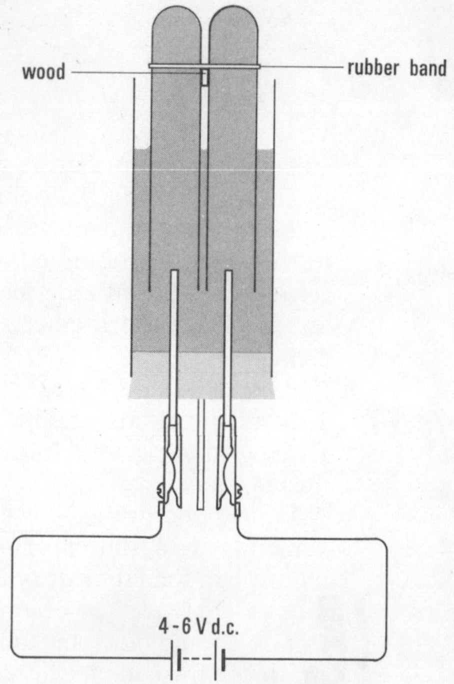

The two small test-tubes are used for collecting any gases given off, and they are held together by a rubber band which has a small strip of wood beneath it. The wood rests on the top of the

Experiment sheet 31
(continued)

wider glass tube, preventing the small test-tubes from covering the electrodes completely. This is important – can you explain why?

Pour the solution to be investigated into the wide tube so that the carbon electrodes are just covered. Fill each test-tube with more solution, over a sink, and invert them together – the solution should remain in the tubes. Still over the sink, put the inverted tubes in the wide tube so that one tube is over each electrode. Connect the electrodes to a battery (or other source of direct current), using crocodile clips and wires. Note which electrode is connected to the positive terminal (this electrode is the anode) and which to the negative terminal (the cathode).

Note what happens at each electrode. If a gas is given off collect one test-tube full, disconnect the battery, and carry out suitable tests on the gas, as discussed with your teacher.

Enter all observations and results in the table.

Solution electrolysed	Action at anode	Anode product	Action at cathode	Cathode product
Copper(II) chloride				
Zinc bromide				
Potassium bromide				
Sodium iodide				

Correlate the information gained from the experiments in this and the last section on the board. List all the elements released at the cathode and all those released at the anode. The pupils will see that, in the reactions obtained, copper, zinc, lead, and hydrogen were released at the cathode and chlorine, bromine, and iodine at the anode. The pattern emerging is studied further in Topic B7 'The elements'.

Suggestions for homework

1. Some metals are obtained by the electrolysis of one of their salts. Design an apparatus to make sodium from common salt by this means.
2. Design and draw a piece of apparatus which could be used conveniently as a source of one of the gases oxygen, hydrogen, or chlorine in the laboratory using electrolysis to produce the gas.

Summary

Pupils should now be familiar with a method of collecting the products formed at the electrodes when aqueous solutions are electrolysed, and they should know the names of the products formed at the anode and cathode in the case of the solutions mentioned in Experiment B6.3 (*Experiment sheet* 31).

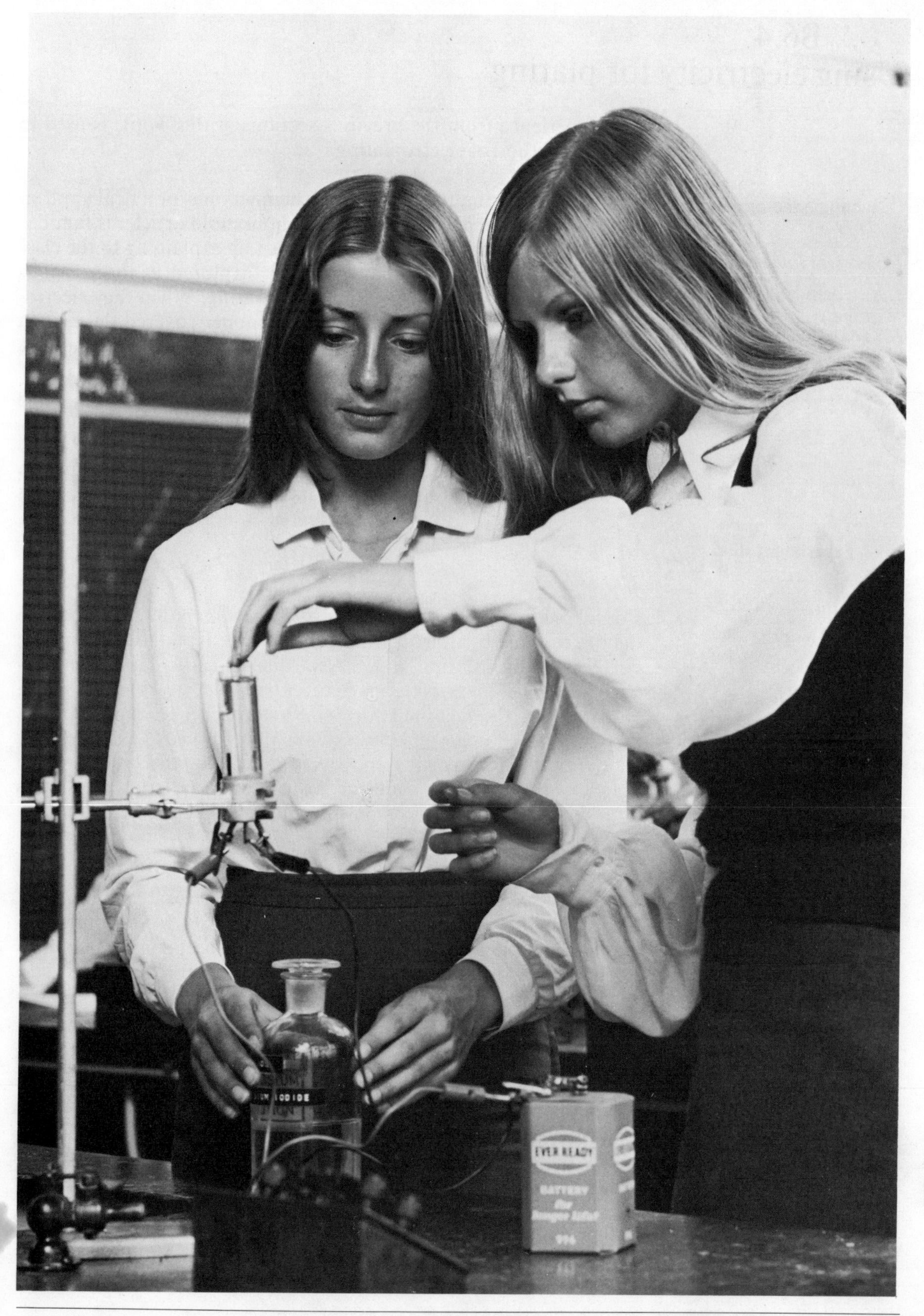
EVER READY
BATTERY

B6.4
Using electricity for plating

The experience from the previous sections in this Topic is used to devise a method of electroplating.

A suggested approach

Objectives for pupils

1. Awareness of the usefulness of chemistry
2. Knowledge of the industrial process of electroplating

This is an opportunity to show that chemistry has practical applications. Chromium plating on cars and household articles is familiar to most boys and girls, and you may start by explaining to the class that the same principles that they have explored in the previous sections are used in industry for electroplating. When they electrolysed copper(II) chloride, they saw that copper covered the carbon rod; in a similar way other metals can be made to cover metallic articles. They may do a simple 'electroplating without a battery' experiment by dipping their penknives or other steel articles into a copper sulphate solution. The blade becomes coated with copper. Explain, after this, that most metals do not plate in this way as electrical energy is needed to make the reaction go. Let the pupils do their own electroplating experiments. An example is given below.

Experiment B6.4

Apparatus

Each pupil (or pair) will need:

Experiment sheet 32

6 V battery or alternative d.c. supply

Bulb and bulb holder

Copper foil cathode, 5 cm × 3 cm approximately

Zinc foil anode, 5 cm × 3 cm approximately

3 lengths of connecting wire fitted with crocodile clips

Support for electrode foils

Beaker, 100 cm^3

Steel wool

Paper tissues

Plating solution, about 100 cm^3 each

M sodium hydroxide

Distilled water

Using electricity for zinc plating

Procedure

For all plating experiments the cathode on which the metal is to be deposited must be thoroughly clean and free from grease, and the correct plating solution and current density used. The circuit has been designed to give the correct current density provided that electrodes of the recommended size are used. The bulb indicates that current is flowing and limits this so that no rheostat is necessary. Immerse 10 to 15 cm^2 of the electrode. Clean the objects to be plated by rubbing them with steel wool and paper tissues moistened with molar sodium hydroxide solution, and then wash them well with distilled water.

A solution suitable for zinc plating may be made as follows: add an equal volume of water to about 50 cm^3 of a saturated solution of zinc sulphate. To this solution add a few drops of M sulphuric acid and two spatula measures of boric acid. A current density of 0.01 to 0.03 amp per cm^2 gives good results in five to ten minutes. Reverse the sides of the cathode when half the time has elapsed to obtain even plating.

Further details are given in *Experiment sheet* 32 reproduced below.

Experiment sheet 32

The discoveries made in the last few experiments can be used as a method for trying to deposit a coat of one metal on another.

The object to be plated (a piece of sheet copper is convenient to use for practice) is made the *cathode* and a sheet of the metal to be plated (zinc in this experiment) is made the *anode*.

The electrolyte in the beaker is a solution containing a salt of zinc.

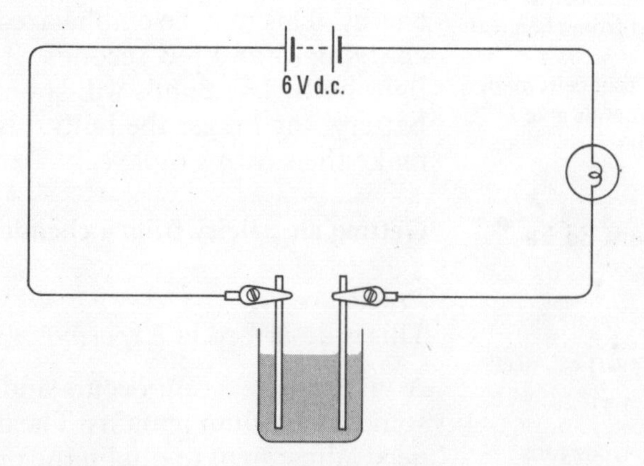

Clean the copper sheet (on both sides) by scouring with steel wool, followed by rubbing with a tissue moistened with dilute sodium hydroxide solution. **Do not let this solution get on to your skin – if by accident it does, wash it off *immediately* with a stream of water from the tap.** Finally wash the copper sheet with distilled water.

Nearly fill the beaker with the zinc plating solution. Support the zinc electrode (anode) and the cleaned copper sheet (cathode) in the beaker and connect them into the circuit shown in the diagram. Between 10 and 15 cm^2 of the copper sheet must be in the plating solution. Allow electrolysis to proceed for five to ten minutes. Remove the copper sheet, wash with distilled water and dry in warm air. What happens?

Suggestion for homework

In your home find as many articles as you can which have been electroplated. Bring back a list of the articles with a description of the type of plating used (they will probably all be chromium or silver plated).

Summary

Pupils should now be familiar with the use of electricity for electroplating.

B6.5
Electricity from chemical reactions

This section is intended to show that electricity can be obtained from chemical reactions, just as electricity can be used to make chemical reactions go. The pupils learn that the potential difference between two metals in solution varies with the type of metal.

A suggested approach

In the case of electrolysis, electrical energy was used to cause a chemical change. Can a chemical change be used to produce electricity? In this lesson you may start by discussing this problem. In

Objectives for pupils

1. Awareness that electricity can be obtained from chemical reactions
2. Recognition that cells made from different metals give different voltages

fact the pupils will all be familiar with batteries for torches and cars. Explain that a battery is a chemical means of storing electrical energy. This may be emphasized by taking a 'flat' accumulator and charging it for a few seconds. It may then be discharged through a light bulb. The pupils will soon see that the longer you charge the battery, the longer the bulb stays bright. Now let the pupils try to make their own simple cell. Details are given below:

Experiment B6.5a

Getting electricity from a chemical reaction

Apparatus

Each pupil (or pair) will need:

Experiment sheet 33

Support for electrode foils

1.25 V at 0.25 A bulb and holder

2 lengths of connecting wire fitted with crocodile clips

Beaker, 100 cm^3

M or 0.5M sulphuric acid

Copper foil, 5 cm × 3 cm approximately

Magnesium ribbon, 15–20 cm approximately

Zinc foil, approximately 5 cm × 3 cm

The copper foil and magnesium ribbon should be cleaned thoroughly using sandpaper

Procedure

This is described in *Experiment sheet* 33 reproduced below.

A vigorous reaction occurs and the bulb glows steadily as long as some magnesium remains. The distance between the electrodes may need adjustment to obtain the maximum illumination.

Experiment sheet 33

So far you have been using an electric current to bring about changes in solutions. It is interesting to see if you can reverse the process and use chemical reactions as a method of producing electric currents.

Set up an 'electrolytic cell' with a piece of copper foil and a piece of magnesium ribbon as the two electrodes. Use dilute sulphuric acid as the electrolyte. Connect the metals by means of copper wires to a torch bulb. Then dip the metals in the acid. Does the lamp light up?
What else do you see happening?

Set up another cell like the first one, but this time use the copper plate again and a zinc plate for the other electrode. Try the effect of this pair on the lamp. What happens?

After the pupils have done this experiment discuss the importance of batteries and lead to the facts that different batteries (*a*) use different metals, and (*b*) produce different voltages. A lead accumulator and a nickel–iron cell or carbon–zinc torch battery may be quoted as examples. Which metals make the best batteries? It is possible to measure the potential difference set up between two different metals very simply by means of the following experiment.

Experiment B6.5b

Comparing the voltages from different cells

Apparatus

The teacher will need:

Projection microammeter and leads fitted with crocodile clips

Projector

(Continued)

A projection microammeter, range 250–0–250 μA and low internal resistance, may be used in series with a suitable resistor to give direct e.m.f. values. The instrument may be mounted in a film strip-slide projector or placed on an overhead projector and the experimental results projected to the class. With a projection microammeter of internal resistance of about 30–50 Ω, a series resistor of 10 000 Ω is needed for a deflection of 200 μA ≡ 2 V, or 5000 Ω for a deflection of 200 μA ≡ 1 V, neglecting the resistance of the instrument.

Beaker, 250 cm^3

Metal foils or strips: copper, zinc, iron, magnesium, silver, nickel, lead, gold

Should a projection microammeter not be available a large demonstration instrument could be substituted. However, if portable multirange meters are available the pupils could record their own results.

Procedure

Copper or iron may be selected as the reference metal. Other metal strips are placed in turn in the beaker containing tap water, to which a little sodium chloride or potassium nitrate may be added. The projection microammeter with its resistor is then connected in circuit. The e.m.f. of the pairs of electrodes should be projected in a vertical plane on the board and the needle position marked for each metal. A list of metals will thus be drawn up on the board in order of the e.m.f. that they produce when coupled to the copper or iron electrode.

The result of this experiment shows that the metals can be arranged in order according to the voltage that they give when compared with copper or iron. This order may be seen to correspond with their general reactivity. The pupils will have seen that some metals burn in oxygen more vigorously than others, and they may be able to link this reactivity with the table they have just seen produced. Too much emphasis should not be placed on 'vigour of reaction' as this, of course, depends on many factors such as the state of division of the metals, which can give misleading impressions of the reactivity of the metal.

Suggestions for homework

1. You have obtained some electricity from a chemical reaction. Find out about similar sources of electricity which are used in the home. You might take apart a run-down torch cell to see how it is constructed.
2. Make a short summary of the work you have done in this Topic.

Summary

Pupils should now know that electricity can be obtained from chemical reactions, and that different pairs of metals in a cell produce different voltages.

Topic **B7**

The elements

Purposes of the Topic

To discuss the meaning of the term 'element' using the experience of chemical reactions that the pupils have had up to this point.

Contents

B7.1 What is an element?

Timing

One single or one double period should be sufficient for this Topic.

Introduction to the Topic

The object of this Topic is to discuss the meaning of the term 'element'. By the end of this Topic the pupils should know what is meant by an element, how a chemist finds out whether a substance is an element or not, and which of the substances they have met so far are in fact elements. They should know that elements can be divided into metals and non-metals, and that in electrolysis metals (and hydrogen) are released at the cathode and non-metals at the anode. This approach is amplified in the *Handbook for teachers*, Chapter 3, pages 56–59.

Alternative approach

An alternative approach to this subject involving some practical work on the properties of elements, and a simple classification of them, is given in Alternative A, Topic A5. This alternative does not depend upon the electrolysis Topic (A8), and can be done without any previous knowledge of electrolysis.

Background knowledge

Pupils already understand what is meant by a single substance, and have some idea of the difference between a mixture and a 'compound' largely from the different methods of separation required: distillation, chromatography, and so on for separating mixtures, and electricity or heat for separating compounds. This knowledge must now be built on, so as to establish the idea of elements as the ultimate product of all separation techniques. Although pupils have not heard the word 'element' so far, they have met the following elements in earlier sections of Alternative B:

Element	Section	Element	Section
Bromine	6.3	Mercury	6.1
Chlorine	6.3	Nickel	6.5
Copper	5.1 6.3 6.4 6.5	Nitrogen	3.3 4.1
Gold	6.5	Oxygen	3.3 4.1
Hydrogen	6.3	Silver	6.5
Iodine	6.3	Sulphur	6.1 6.2
Lead	6.1 6.3 6.5	Zinc	6.3 6.4 6.5
Magnesium	6.5		

Subsequent development

The theme of elements and their properties is continued in Topic B8 'Further reactions between elements' and Topic B10 'Competition among the elements'.

Reading material

for the pupil

Study sheet:
The chemical elements. The theme of this is how the idea of the elements has developed in history with the discoveries by such famous scientists as Boyle, Davy, Mendeleev, and Ramsay. In describing this development, the Study sheet aims to help pupils acquire the idea of what an element is.

B7.1
What is an element?

A discussion of the concept of an element using the experience of chemical reactions that the pupils have had up to this point.

A suggested approach

Objectives for pupils

1. Development of the chemist's concept of an element
2. Understanding of the meaning of the word element

Enough evidence has now been amassed to discuss the meaning of the term 'element'. This subject is perhaps best introduced historically. You may explain that the theory of elements is that substances are made up of other simpler, elementary, substances. The problem then is what these fundamental elements are. The Greeks thought that they were fire, air, water, and earth. A very reasonable scheme was worked out on this theory. This was accepted until Boyle made the modern proposal that one should only consider a substance to be an element if it could not be broken up into any other substances.

To find out whether any of the substances we have used is an element we must try to see if it can be broken up. How can we break it up? By heat – by electricity. We have tried the effect of both on many substances. What was the result? At this point it might be worth taking some examples and discussing them. Was the substance changed by heating it in air? What happened when an electric current was passed through it? Did it decompose? Was the substance being considered one of the products of an electrolysis?

When all these questions are answered we may be able to say that 'it looks as though . . .' a particular substance is an element. Explain that until comparatively recently there were no absolute methods of telling whether a substance was an element or not, and for a long time chemists had to use the sort of argument that the pupils have used in the lesson. Having discussed the theory of elements you may go on to talk about ways of dividing them into groups. The pupils will have noticed that the elements which were released at the cathode were all (except for hydrogen) similar. They were all metallic. Those released at the anode were not metallic. Explain that this is a fundamental division and that elements can be divided into metals and non-metals. Unfortunately nature does not like tidy boundaries and there are some elements that sit on the fence and show the properties of both metals and non-metals.

New word

Element

Suggestions for homework

1. Read the Study sheet *The chemical elements*.
2. Choose any one element that interests you and find out as much as you can about when and by whom it was discovered, from where it is obtained, and for what it is used.

Summary

Pupils should now understand that an element is a substance that cannot be broken up into any other substances (for example, by heat or electricity). They should know that elements can be divided roughly into metals and non-metals.

Topic **B8**

Further reactions between elements

Purposes of the Topic

The main purpose of this Topic is for the pupils to examine the properties of some elements and their compounds, so as to get some idea of the chemical differences between elements and compounds. It is not intended that pupils should learn a set of dictated 'differences' but rather that they should get the general ideas of the differences through familiarity with the behaviour of some materials.

It is also intended that pupils will see that elements with similar properties tend to form similar types of compounds.

Contents

B8.1 Studying the difference in properties between elements and their compounds: hydrogen, oxygen, and water
B8.2 Studying the difference in properties between elements and their compounds: some sulphides, oxides, and chlorides

Timing

Section B8.1 will need at least two double periods and section B8.2 will probably need three. A total of six double periods should give plenty of time for the Topic as a whole.

Introduction to the Topic

There are two distinct parts to this Topic. In the first part, section 8.1, we are concerned with the differences in properties between the elements hydrogen and oxygen, and their compound water. Having established this difference in properties, which can be emphasized through the rest of Alternative B, Stage I, we look at groups of compounds and see if there are ways in which these groups of compounds are related to the elements from which they are formed. The examples chosen are the sulphides, oxides, and chlorides; the properties chosen are those of solubility and pH in solution. The pupils find that these properties, pH in particular, are related to the metallic or non-metallic nature of the elements.

By the end of this Topic, therefore, the pupils should know through the experiments they have performed that compounds have chemical and physical properties which are quite different from those of the elements of which they are composed. They should know also that elements which are chemically related tend to form compounds which are also chemically related. The acidic nature of solutions of the oxides of non-metals and the alkaline nature of the soluble oxides of metals are examples of this.

Background knowledge

This Topic leads on directly from the preceding Topic, in which the idea of elements was established.

Subsequent development

The ideas developed in this Topic, and a number of the elements and compounds met here, are used again in Topic B10 'Competition among the elements'.

B8.1
Studying the difference in properties between elements and their compounds: hydrogen, oxygen, and water

The differences between some elements and their compounds are discussed in the light of previous experience. The elements hydrogen and oxygen are studied, together with their compound water.

A suggested approach

Objectives for pupils

1. Use of the ideas of elements and compounds
2. Knowledge of some simple properties of hydrogen, oxygen, and water

Start with a discussion of some of the compounds which the pupils have already made. In Topic B4 they burned a number of elements in oxygen. The differences between the elements they started with and the compounds which these elements produced were very marked. Magnesium, a bright metal, and oxygen, a colourless gas, produced magnesium oxide, a white powder. You might ask them for examples which show a similarly large difference in properties. After some discussion along these lines, introduce the elements for investigation in this Topic. These are hydrogen and oxygen. Some preliminary experiments may be done to confirm some of the properties of the gases which they discovered during Topic B6. The main questions to be answered are 'Are the gases heavier or lighter than air?', 'Are they soluble in water?', and 'How can we identify them?' A series of simple experiments is described below.

Experiment B8.1a

Properties of hydrogen and oxygen

Apparatus

Each pupil (or pair) will need:

4 test-tubes, 150 × 25 mm, fitted with corks

Test-tube rack

Shallow beaker or small trough

Wood splint

Indicator paper

Access to:

Cylinders of hydrogen and oxygen

Procedure

Fill three test-tubes with hydrogen from whatever source is available and cork them up. One tube is opened under water in the beaker or trough to find out whether the gas is soluble. The second is held *beneath* an empty test-tube for a few seconds so that the hydrogen passes from the lower to the upper test-tube. A light is applied to both to find out where the hydrogen is. A similar experiment is performed with the third test-tube of hydrogen but it is held *above* an empty test-tube. A lighted splint applied to the mouth of each shows that the hydrogen has not passed downwards to the lower test-tube. Be careful not to hold the test-tubes together for too long as some hydrogen may enter the lower tube by diffusion.

Three test-tubes are then filled with oxygen, and similar experiments on solubility and density performed. The presence of the oxygen should be detected by the use of a glowing splint instead of a flame.

There is no *Experiment sheet* for this experiment; it is suggested that the teacher should give the pupils their instructions; alternatively he may care to demonstrate the experiment.

After this experiment, turn the attention of the class to the properties of the compound of hydrogen and oxygen, namely water.

Hydrogen can be burnt in air as explained in the next experiment, and the properties of the water formed can then be examined.

Experiment B8.1b

Making water from hydrogen and oxygen

This experiment **must** be performed by the teacher.

Apparatus

The teacher will need:

Plastic safety screen

U-tube or calcium chloride tube

Thistle funnel with stem bent as in diagram, fitted with bung for the test-tube

Test-tube with side arm, 125 × 16 mm

Beaker, 250 cm^3

2 stands and clamps

Filter pump and connection tubing

Metal or ceramic jet and connection tubing

Test-tube, 100 × 16 mm

Thermometer, −10 to +110 °C

Bunsen burner and asbestos square

Electrodes, 6 V bulb, and battery for testing conductivity

3 lengths of connecting wire fitted with crocodile clips

Hydrogen cylinder

Silica gel or anhydrous calcium chloride

Anhydrous copper(II) sulphate

Procedure

The hydrogen is first dried by passing it through a tube containing silica gel or anhydrous calcium chloride. When all air has been displaced completely from the apparatus the hydrogen is burnt at a metal or ceramic jet. The hydrogen flame should be about 3 cm high and be allowed to burn beneath the thistle funnel as indicated in the diagram. The gas formed by the combustion is drawn through the apparatus by a filter pump, and a liquid soon condenses in the cooled test-tube. After about ten minutes 2 or 3 cm^3 of liquid will have collected. Test the product for conductivity with a pair of electrodes, bulb, and battery. The liquid may be confirmed to be water by (*a*) testing it with anhydrous copper sulphate, or (*b*) determining its boiling point.

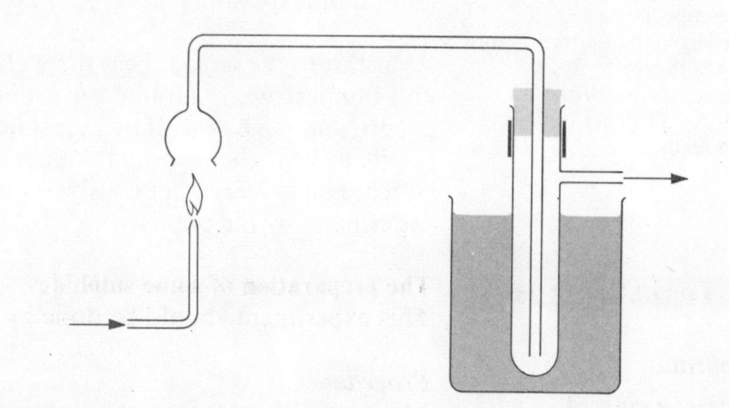

Discuss the results with the class, and summarize the differences between liquid water and the gases hydrogen and oxygen on the board.

Suggestion for homework

Select five compounds and show how these compounds differ from the elements of which they are composed.

Summary

Pupils have studied the elements hydrogen and oxygen and their compound water. They should have noted from this example that the compound is made up of the elements but has properties of its own, quite distinct from those of the elements (for example in physical state, density, solubility, combustibility, action on anhydrous copper sulphate, and so on).

B8.2
Studying the difference in properties between elements and their compounds: some sulphides, oxides, and chlorides

A number of chlorides, sulphides, and oxides are studied. The pupils see a marked difference between the metal and non-metal compounds, and they are led to find a gradation in the properties of the metal compounds.

A suggested approach

Objectives for pupils

1. Use of the ideas of elements and compounds
2. Recognition of a relationship between the reactivity of elements and the degree of acidity or alkalinity of their compounds

Only a few of the compounds whose properties are to be studied in this section should be prepared. When the pupils have seen that the direct preparation of these binary compounds is comparatively simple they should be given samples of a number of these compounds for study.

As a start to the lesson, heat three elements, for example, zinc, iron, and copper, with sulphur on asbestos paper. For details of this experiment see below. If there is time, the teacher may like to get the pupils to heat the iron and copper with sulphur themselves. Zinc, which reacts very vigorously with sulphur, is not suitable for experiment by the pupils.

Experiment B8.2a

Apparatus

The teacher will need:

Tongs

Spatula

Pieces of asbestos paper, 7 cm × 2 cm approximately

Sulphur powder

Zinc dust

Iron powder

Copper powder

The preparation of some sulphides

This experiment should be done by the teacher.

Procedure

Mix a spatula measure of sulphur with a spatula measure of each metal powder in turn. Heat a small quantity of each mixture on a piece of asbestos paper, folded so that its cross-section is V-shaped, and hold the asbestos paper with a pair of tongs. Point out the evolution of heat in each case; the reaction with the zinc is the most vigorous.

As a further illustration of the formation of binary compounds the teacher may now demonstrate the formation of chlorides by burning several elements in chlorine as described below.

Experiment B8.2b

Apparatus

The teacher will need:

Combustion spoon suitable for use in gas-jars

Tongs

Bunsen burner and asbestos square

(Continued)

The preparation of some chlorides

This experiment should be done by the teacher.

Procedure

Heat each element in turn and plunge it into a jar of chlorine. Use very small quantities, particularly of the sodium and phosphorus, the size of which should be smaller than a rice grain. The steel wool can be held in tongs, heated to redness and at once transferred to the chlorine. The Dutch metal should be dry and need not be heated beforehand. The sodium and phosphorus should be melted in a combustion spoon and then held in the chlorine. In each case point

4 gas-jars of chlorine

Steel wool

Sodium

'Dutch metal'

Phosphorus

out the evolution of heat as the element combines with chlorine. Note that the sodium gives a coloured product owing to attack by the chlorine on the spoon.

Having shown that a number of compounds can be made by direct combination, and that energy is evolved when this happens, you may now introduce a larger range of compounds for study. There are several ways of handling this. For instance, each pair might be given a different compound and all the results collected together at the end of the lesson. Alternatively, each pair might be given a series of compounds – for example, the chlorides, sulphides, or oxides – to study, and the results again collected at the end of the lesson. In any case the lesson will need a good deal of organization. Some teachers might like to select a smaller number of substances for testing. The main questions to be asked are: 'What is the compound like?' (that is, state, colour), 'Is heat evolved when the compound is dissolved in water?' and 'What is the pH of its solution in water and does its solution in water conduct electricity?' The pupils should then carry out the experiment of which details are given below.

Experiment B8.2c

Some properties of oxides, chlorides, and sulphides

Apparatus

Each pupil (or pair) will need:

Experiment sheet 34

6 test-tubes, 100 × 16 mm, in a rack

Full-range Indicator paper

Electrode holder with pair of steel electrodes

6 V battery or alternative d.c. supply

6 V bulb in holder

3 lengths of connecting wire fitted with crocodile clips

Distilled water

Chlorides: sodium, magnesium, zinc, iron(III), and copper(II) chlorides; phosphorus trichloride; and tetrachloromethane

Sulphides: zinc, iron(II), copper(II), sodium, and calcium sulphides; carbon disulphide

Oxides: magnesium, zinc, iron (III), calcium, and copper(II) oxides;
phosphorus pentoxide

Note. The pupils should not be allowed access to the bottles of tetrachloromethane, phosphorus trichloride, and carbon disulsulphide, as these compounds are too poisonous or dangerous for them to handle at this stage. If he likes, the teacher can give some pupils one drop of each of the liquids in test-tubes, but no more.

Procedure
The class should test for solubility noting any heat change on addition of water, and test the solution for pH and for electrical conductivity, with very small quantities; one drop in the case of the liquids, and half a spatula measure in the case of solids, adding distilled water to half fill the test-tube.

Experiment sheet 34
You have seen how oxides, chlorides, and sulphides of some elements can be made. In this experiment you will investigate the properties of some of these compounds.

A range of compounds will be provided. For each, perform the following tests:
1. Investigate the solubility in water by shaking a little of the compound with about one-third of a test-tube of distilled water. Then add further small portions until no more dissolves. Is the compound insoluble, slightly soluble, or very soluble? Is heat evolved or absorbed by reaction with water?
Enter your answers in the table (overleaf).

Experiment sheet 34
(continued)

Compound	Solubility in water	Heat evolved or absorbed	pH of solution	Electrical conductance

2. Test a drop of the solution with Full-range Indicator paper. Record the pH.
3. Find whether the solution conducts electricity.

Now hold a discussion to collect the results together. The pupils will have seen that the metal oxides, sulphides, and chlorides are solids but that the non-metal compounds (except phosphorus pentoxide) are liquids. If the metals are put in order of their reactivity, the pupils will see a gradation in the properties of each type of compound. Solutions of sulphides and oxides become more alkaline as the metals become more reactive and solutions of chlorides become less acidic. It should be brought out during this discussion that the properties of a compound depend on *both* the constituent elements. It is essential to give plenty of time to this post-experiment discussion as a large number of experiments will have been performed, leaving the pupils with a mass of uncorrelated facts which must all be pulled together by the teacher.

Suggestion for homework

Make a scheme of all the facts you have discovered in this Topic and try to find patterns of behaviour among the elements and their compounds.

Summary

Pupils should now have a fairly clear idea of the relationship between an element and a compound, and some knowledge of a number of oxides, sulphides, and chlorides – all binary compounds. They should be aware of certain trends in the properties of compounds: solutions of sulphides and oxides become more alkaline as the metal becomes more reactive, and solutions of chlorides become less acidic as the metal becomes more reactive.

Topic B9

Investigation of some common processes involving the air

Purpose of the Topic

To investigate the chemical changes which take place when burning, breathing, and rusting occur.

Contents

Timing

One double period should be allowed for each section except for B9.1 and B9.3, which will probably need two. A maximum of seven double periods for the whole Topic may be allowed to give time for additional discussion of experiments, if necessary. If time is short, sections B9.3 and B9.4 could be condensed into one double period by leaving out most of the experimental work. The Topic could then be completed in three double periods.

Introduction to the Topic

This Topic is essentially an investigation into the nature of burning, breathing, and rusting – all phenomena which involve the air. By the end of this Topic the pupils should understand that burning is a reaction with oxygen. In examining this further they should have found that some fuels and foods give carbon dioxide and water on combustion, indicating the presence of carbon and hydrogen in the fuels or foods. Analysis of exhaled air reveals also the presence of carbon dioxide and water and this leads to the idea that the food we eat undergoes in our bodies a series of reactions similar to that of combustion. They should know that these reactions are the principal source of energy for the body. From the investigation on rusting they should have learnt that water and air are both necessary for rusting to take place and that rust must contain oxygen and hydrogen.

Alternative approach

An alternative approach to the chemistry of burning is to be found in Alternative A, Topics A3 and A4. In Alternative A the investigation of burning leads to the discovery of the composition of the air.

Here the composition of the air is already known (Topics B3 and B4) and so the investigation is mainly centered around the products of burning. Breathing and rusting are not considered in Alternative A.

Background knowledge

Pupils already know, from Topic B4, that the air consists mainly of oxygen and nitrogen, and that substances burn in oxygen but not in nitrogen.

Supplementary materials

Films
BBC programme 'Corrosion'
BBC programme 'Burning'
See Appendix 4 for brief descriptions and further details

Reading material
for the pupil

Study sheet:
Burning and Lavoisier starts with early ideas about burning, including those of the phlogistonists. Lavoisier's great contribution to our understanding of the processes of combustion and breathing is simply explained and set in the context of his life and times.

B9.1
What chemical changes take place in burning?

In this section the pupils investigate the chemical changes which take place when a number of everyday substances are burned.

A suggested approach

Objectives for pupils

1. Knowledge that carbon dioxide and water are formed when food and fuels are burnt
2. Ability to test for carbon dioxide and water
3. Understanding of the meaning of the term ignition point

Useful film

BBC programme 'Burning'

See Appendix 4 for a brief description and further details

Start by reminding the pupils of what they learnt in Topic B4 'The major gases of the air'. At that time they burnt a number of elements in air and oxygen. To investigate the matter of burning further they may burn some everyday substances such as paper, wood, coal gas, candle wax, sugar, starch, polythene, ethanol, and naphthalene. The last two are less well known, but the pupils have come across them in the early part of this course. Distribute the materials around the class so that the groups burn different substances. Now get them to burn the substances and test for the products. In all these cases, water and carbon dioxide will be formed. Heat on a combustion spoon, place in a test-tube of air, and test the products with limewater and cobalt chloride paper. Use a teat pipette to extract the carbon dioxide produced by the burning solid and to bubble it through the limewater.

Ask the pupils to note the relative amount of heat given off by each substance when it is burnt. Draw the pupils' attention to the fact that energy is needed to 'trigger off the reaction' and introduce the term: ignition point.

Experiment B9.1

An investigation of some of the chemical changes that take place during burning

Apparatus

Each pupil (or pair) will need:

Experiment sheet 35

4 test-tubes, 150 × 25 mm, fitted with corks

Combustion spoon

Bunsen burner and asbestos square

(Continued)

Procedure

Let the pupils take wood, paper, polythene, naphthalene, sugar, starch, ethanol, and a portion of wax taper, and in turn ignite them on a combustion spoon, placing the spoon in a dry 150 × 25 mm test-tube containing air. Sugar and starch require strong heating before they will ignite.

As soon as the flame is extinguished, they should withdraw the spoon and cork the test-tube.

A mist will be noticed on the inside of the test-tube, indicating that water is formed, and the addition of limewater after burning will

Paper

Wood

Naphthalene

Sugar

Polythene

Starch

Ethanol (industrial methylated spirit)

Wax taper

Limewater

The teacher will need:

Length of rubber tubing

15 cm length of glass tubing drawn out to a jet, or glass dropper from a teat pipette

Round-bottom flask, 250 cm^3

show that carbon dioxide is also present. Details are given in *Experiment sheet* 35 reproduced below.

Experiment by the teacher

Take a piece of glass tubing drawn out to a jet and connect it to the gas supply by means of a piece of rubber tubing. The small gas flame thus obtained can be burnt in a round-bottom flask containing air, as shown in the diagram. As soon as the flame goes out switch off the gas supply and withdraw the glass tubing. Test the products of combustion as before.

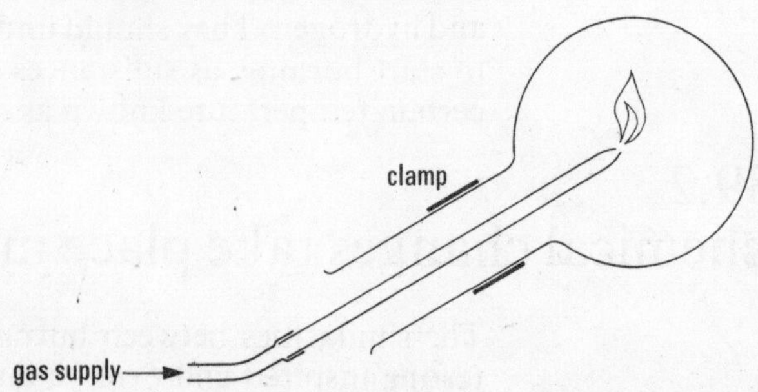

Warning. If the experiment is repeated, make sure that there is no residual gas in the flask before putting the lighted jet in it.

Experiment sheet 35

You are going to burn a number of substances in large test-tubes filled with air, so that you can test for the products of combustion.

Put a little of one of the substances in a combustion spoon, heat in a Bunsen flame until it begins to burn and lower the spoon into a large dry test-tube. When the burning stops, remove the spoon, and cork the test-tube.
What do you notice on the walls of the tube?
Remove the cork, add a little limewater, replace the cork, and shake the tube. What happens?
What does this indicate?

Repeat the process with the other substances which you are given. Enter the results in the table.

Substance	Appearance of walls of tube	Action with limewater	Products of burning

The discussion of the results after the experiment should bring out the fact that if carbon dioxide is present as a product of combustion the material burnt must have contained carbon. Similarly, if water is formed, the substance must have contained hydrogen. It looks, from the experiment, as though a number of substances used as fuels and foods contain hydrogen and carbon.

New word

Ignition point

Suggestion for homework

Write an essay on the importance of fuels in everyday life. Why do you think some fuels are preferred to others? What are the advantages of wood, coal, oil, and coal gas as fuels? Which fuels cause air pollution?

Summary

Pupils should now know that burning is a chemical reaction with oxygen. They have seen that a number of foods and fuels when burned give carbon dioxide and water, and so must contain carbon and hydrogen. They should understand that energy is often required to start burning, as substances only begin to react with oxygen at a certain temperature known as the ignition point.

B9.2
What chemical changes take place in breathing?

The similarities between burning and breathing are discovered by testing inspired and expired air.

A suggested approach

Objectives for pupils

1. Knowledge of the chemical changes occurring in breathing
2. Ability to test for carbon dioxide and water

Start by discussing breathing. It is clear that air is taken into the lungs and then expelled. What happens to it in this time? We can approach this qualitatively at first. What is the difference between the inspired and expired air? The pupils may test expired air for moisture and carbon dioxide and compare it with the air in the laboratory. Having established that carbon dioxide and water vapour are present in expired air in greater proportion than in inspired air, you may ask where they have come from. In the last lesson the pupils saw that many compounds, including starch and sugar, burned giving carbon dioxide and water. This will lead them to the idea that burning and breathing are similar processes. Point out that just as energy was given off when starch was burnt, it is given off when starch is 'burnt' in the body.

Experiment B9.2

Apparatus

Each pupil (or pair) will need:

Experiment sheet 36

2 test-tubes, 100 × 16 mm

Side-arm test-tube with cork or bung and glass delivery tube (see diagram)

Filter pump and connecting tubing

Thermometer, −10 to +110 °C

Beaker, 100 cm^3

(Continued)

An investigation of breathing

Procedure

1. For this experiment the pupil breathes out slowly and steadily through a piece of glass tubing or a drinking straw into some limewater in a 100 × 16 mm test-tube, and notes the time required for the solution to turn milky (this may be about ten seconds).

The pupil then draws air through some limewater using a filter pump (see diagram opposite). The volume of limewater should be the same as that used in the breathing experiment, and the air should be drawn through at the same rate. The length of time which it takes for the milkiness to appear is noted (this may be about ten minutes). The filter pump is left on for the next experiment.

2. Carry out a similar pair of experiments with empty and dry test-tubes, this time placed in ice-cold water. Maintaining the same gas flow rates, the pupil should breathe into one tube and draw air

Glass tube (about 0.4 cm o.d. and 25 cm long) or drinking straws

Stand and clamp

Ice

Limewater

Anhydrous copper sulphate

through the other and record the times taken for the condensation of detectable amounts of liquid in each case. The liquid is identified as water by the addition of anhydrous copper sulphate.

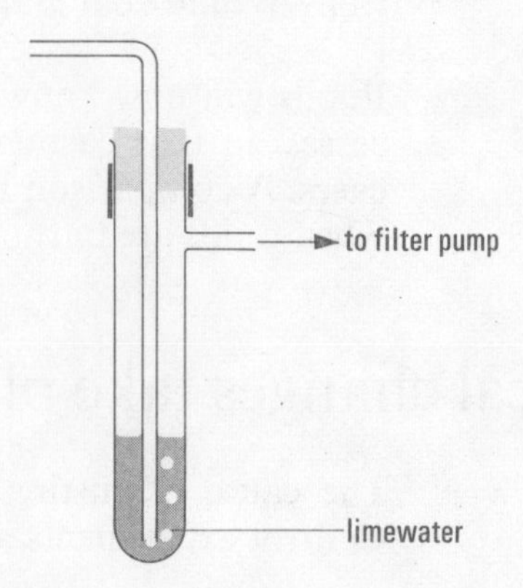

3. To find out about the difference in temperature between inspired and expired air, a thermometer should be read while in ordinary air, and then again while the pupil breathes on to the bulb gently, the bulb being held about 5 cm from the mouth. Body temperature may also be recorded from the closed palm of the hand.

Experiment sheet 36 is reproduced below.

Experiment sheet 36

You are going to compare some of the properties of air before and after it has passed through your lungs.

1. Breathe gently through a piece of glass tubing into some limewater in a test-tube. The time required for the limewater to turn cloudy is . . .

Using a water pump, draw air, at about the same rate as in the 'breathing' experiment above, through a similar volume of limewater. The time required for the limewater to turn cloudy is . . .

2. Compare the effects of passing (*a*) your breath, and (*b*) air, through dry test-tubes cooled in ice water. What is the difference?
What do you think formed in the test-tube?
Plan a test to confirm your answer, carry it out and describe what happens.
Was your answer confirmed?

3. Breathe gently on a thermometer bulb. Is your breath warmer or cooler than the surrounding air?

Make a brief summary of the differences between inspired and expired air.

Suggestions for homework

1. Read the Study sheet *Burning and Lavoisier*.
2. Which elements seem to be present in foodstuffs? If these elements are present in other materials, why can't we eat them as well? Would coal make a good food?

Summary

Pupils will now know that burning and breathing are similar processes, in that carbon dioxide and water can be formed in both cases. A comparison has been drawn between the energy given out when foods are burnt in air, and when they are 'burnt' in the body.

B9.3
What chemical changes take place in rusting?

The cause of rusting is investigated in a series of experiments. Control experiments are used for the first time.

A suggested approach

Objectives for pupils

1. Ability to design simple experiments
2. Knowledge of the chemical changes occurring in rusting
3. Awareness of the reasons for control experiments

Useful film

BBC programme 'Corrosion'

See Appendix 4 for a brief description and further details

Everyone in the class will be familiar with rusting. They have seen rust on their bicycles, on cars, and on household and garden articles. The first question that you may ask them is 'What causes rust?' Have they noticed whether there are particular conditions under which things rust more quickly? You may remind them at this point of their experiments in B4.1. They may have noticed that their bicycles are more likely to rust if they are left outside. Things rust more readily in the bathroom or the kitchen than in the sitting-room. Another question is 'What sort of objects rust?' They should soon come to the conclusion that only iron or steel objects rust (in the normal use of the word). It looks as though water and air may be responsible for rusting – ask the pupils now how they would test their ideas. If air is necessary, is it the oxygen or nitrogen which is active? Is water necessary and, if so, is pure water or water with a dissolved salt in it more corrosive? A discussion on how to carry out the experiments to test these theories should lead to experiments of the following type being performed by the pupils. Note that for the first time control experiments are being used.

Experiment sheet 37 outlines the problem, gives leading questions, but does not describe the experiments in detail. It is reproduced after suggested Experiment 9.3c.

Experiment B9.3

An investigation of the changes which take place when iron rusts
For this experiment each pupil or pair of pupils will need *Experiment sheet* 37. The apparatus and procedure for each part are as follows.

Experiment B9.3a

Part 1: are both air and water necessary for rusting?

Apparatus

Each pupil (or pair) will need:

Experiment sheet 37

(Continued)

Procedure

Take four test-tubes. In the first put two ordinary wire nails followed by distilled water, enough to half cover the nails. The nails are in contact with air and water, and will provide the control experiment. Then place a cork loosely in the mouth of the test-tube to keep out dust and prevent the water from evaporating unduly.

4 test-tubes, 100 × 16 mm, fitted with corks

Test-tube rack

Spatula

Beaker, 100 cm^3

Tripod and gauze

Bunsen burner and asbestos square

Vaseline or olive oil

8 iron nails, about 3 cm long

Calcium chloride, anhydrous, small lumps

Cotton wool

Salt water

Distilled water

In the second tube place a few pieces of anhydrous calcium chloride, followed by a small plug of cotton wool and then two nails, finally corking the test-tube firmly. These nails will be in contact with air but not with water.

In the third tube place enough boiled water to cover the nails completely. The water should have been boiled in a beaker for several minutes to expel dissolved air. Then put the nails in, and place a little Vaseline, or a few drops of olive oil, on to the water by means of a spatula. The Vaseline will melt and form an air-proof layer on the hot water, solidifying as the water cools. Place a cork in the test-tube which now contains nails in contact with water only. The fourth test-tube is like the first, but contains salt water instead of distilled water.

Stand the four test-tubes in a rack and leave for several days, at the end of which time only the 'control' nails and those in contact with salt water will be rusty.

Experiment B9.3b

Part 2: which gas in the air is responsible for rusting?

Apparatus

Each pupil (or pair) will need:

Experiment sheet 37

Test-tube, 150 × 25 mm

Beaker, 100 cm^3

Spatula

Iron filings or steel wool

Wood splint

Procedure

Moisten the inside of the test-tube with water, sprinkle in about a spatula measure of iron filings, and rotate the test-tube horizontally so that the filings spread and adhere to the walls. Alternatively a small plug of moistened iron wool may be inserted in the upper part of the test-tube.

The test-tube is then inverted in a beaker about one-third full of water, using the beaker lip to support the inclined test-tube. The water level inside and outside the tube should be the same and this value noted. The apparatus is then allowed to stand for a few days. During this time any water lost by evaporation should be made good. The iron will rust and the water level will rise inside the tube, then become steady. When the final water level inside the test-tube is noted, once more at atmospheric pressure, it will be apparent that only one-fifth of the air volume has been used up, suggesting that oxygen rather than nitrogen has been involved. The residual gas does not support combustion of a lighted splint, confirming that the gas used up while the iron rusted must have been oxygen.

Experiment B9.3c

Part 3: confirming that oxygen is the gas in the air which is responsible for rusting

Apparatus

Each pupil (or pair) will need:

Experiment sheet 37

2 test-tubes, 100 × 16 mm, fitted with rubber bungs

2 gummed labels

(Continued)

Procedure

Fill a test-tube with oxygen, by displacement of water, label it 'oxygen', and stopper it securely. Similarly, fill another test-tube with nitrogen, and stopper it. Add a measure of iron filings to each tube, taking care to allow as little interchange of the gases with air as possible. Rotate the tubes so that the filings are distributed evenly over the inner walls. Put both tubes mouth downwards, in a beaker of water, and remove the stoppers.

Beaker, 250 cm^3

Iron filings

Distilled water

The teacher will need:

Cylinders of oxygen and nitrogen (delivery tubes attached to both)

The two test-tubes should then be left for a week at the end of which time the filings in contact with oxygen will be found to be rusty, and the water level will have risen considerably. There will be little change in the tube containing nitrogen.

Experiment sheet 37

1. It is common knowledge that rust is formed quickly on iron surfaces which are exposed to air and water. Plan and carry out a set of experiments in which iron (use iron nails) is exposed separately, in test-tubes, to (*a*) dry air, (*b*) air-free water (remember that air dissolves in water), (*c*) air and water together, (*d*) air and salt water. Can you now answer the question 'Are both air and water necessary for rusting?

2. If air takes part in rusting, which of the gases of the air is responsible, one only or all of them? Try to answer this by leaving some moist iron filings coating the inside of a test-tube inverted with the open end under water for several days. About what fraction of the air is used up? Which of the gases in the air is present in this fraction?
Why is it better to use iron filings for this experiment rather than iron nails?

3. Check your conclusions about the last experiment by leaving moist iron filings in separate test-tubes of oxygen and nitrogen for a few days with the open ends of the tubes under water. Do the results support your conclusions?

Write a brief statement of the conditions which you think cause iron to rust.

Suggestion for homework

Rusting is said to cost millions of pounds a year in this country alone. Write down as many instances as you can of damage done by rusting. How do you think this damage could be prevented?

Summary

Pupils should now know that both oxygen and water are necessary for iron to rust. They should understand the importance of setting up a control experiment.

B9.4
What is the chemical composition of rust?

A suggested approach

In the last section the causes of, and conditions for, rusting were investigated. In this section the rust itself is analysed.

It was clear from the experiments in the last section that oxygen and water are in some way concerned in the rusting of iron. The question now is 'What is rust?' You may start by discussing the problem with

Objective for pupils

Knowledge that rust is a compound containing iron, hydrogen, and oxygen.

the class. 'How can we determine what rust is made of?' A simple experiment would be to heat it and see what happens. Details of this experiment are given below.

Experiment B9.4

Investigating the composition of rust

Apparatus

Each pupil (or pair) will need:

Experiment sheet 38

Beaker, 100 cm^3

Watch-glass (hard-glass)

Tripod and gauze

Bunsen burner and asbestos square

Test-tube holder

Test-tube, 100 × 16 mm

Spatula

Cobalt chloride paper or anhydrous copper(II) sulphate

Rust

Procedure

The pupil should add a spatula measure of the dry rust to a dry 100 × 16 mm test-tube and heat it. Drops of water will be seen to condense on the cooler upper parts of the test-tube. This can be tested, using a little anhydrous copper sulphate. Examine the residue and note its appearance.

Experiment sheet 38 is reproduced below.

> **Experiment sheet 38**
> Obtain some rust from a scrap-heap or elsewhere. Wash it with water and dry it on a watch-glass heated on a beaker one-third full of boiling water.
> Examine the dry rust by placing a measure in a dry test-tube and heating it. Is there any change?
> Write down what happens. What conclusion can you draw about the composition of rust?

From the last experiment the pupils will see that if water was produced both hydrogen and oxygen were present in the rust. Had hydrogen alone been given off it could have been ignited. We must therefore conclude that rust contains hydrogen and oxygen. 'What is there in the residue?' This question will have to be left for consideration in the next Topic.

Suggestions for homework

1. How would you find out if aluminium (or another metal) 'rusts'? If it does, what experiments would you do to find out what conditions are necessary for the 'rusting' of aluminium and what the 'rust' is composed of?
2. Make a summary of the work you have done in this Topic.

Summary

Pupils now know that rust, in addition to iron, contains both hydrogen and oxygen.

Topic **B10**

Competition among the elements

Purposes of the Topic

1. To establish a reactivity series of some metals, and another of some non-metals
2. To use these series to attempt to predict certain properties of elements
3. To consider how the principles behind these series are used in the industrial production of some metals and non-metals

Contents

Timing

Sections B10.1 and B10.3 will probably need two double periods each, and B10.5 may need as many as three. The two other sections should need only one double period each. A total of ten double periods for the Topic as a whole allows time for recapitulation and discussion. If time is short B10.2 may be omitted and B10.5 may be condensed into one double period. This Topic could then be completed in five double periods.

Introduction to the Topic

In this Topic the pupils use their experience with experiments earlier in the course to establish a reactivity series from which they make predictions about chemical reactions. They first use the series to find out which metal oxides can be reduced with carbon and hydrogen. They then use it to predict which metals will replace others from their solutions and finally to try a similar experiment with non-metals. This leads to a study of the halogens as a family.

By the end of this Topic the pupils should know how to build up a reactivity series, how to use it, and *what its limitations are.* They should also have a general knowledge of the halogens as a family of elements.

Alternative approach

As far as the metals are concerned, much the same ground is covered, in a rather different manner, in Alternative A, Topic A6. An alternative treatment of the halogens can be found in Topic A10.

Background knowledge

This Topic makes use, in a general way, of most of the preceding Topics, but certain areas are particularly linked with it. They are:
1. The burning of metals in oxygen (magnesium, calcium, and iron in Experiment B4.1)
2. The relative reactivities of the metals towards sulphur (zinc, iron, and copper in Experiment B8.2a)
3. The relative reactivities of the metals towards chlorine (sodium, iron, and a copper–zinc alloy in Experiment B8.2b)
4. The electrode potentials of the metals (Experiment B6.5b)

Further references
for the teacher

Additional experiments on this theme are to be found in *Collected experiments*, Chapter 7 'Reactivity sèries'.

Supplementary materials

Film loops
1–11 'Fluorine manufacture'
1–12 'Fluorine compounds uses'
1–13 'Chlorine manufacture'
1–14 'Chlorine uses'
1–15 'Bromine manufacture'
1–16 'Bromine uses'
1–17 'Iodine manufacture'
1–18 'Iodine uses'

Reading material

Films
'Making iron'
'Treasure trove'
'Study in steel'
'Chlorine'
See Appendix 4 for a brief description and further details of these films

Reading material
for the pupil

Study sheets:
Competitions. The word 'competition' is one of several whose meaning in chemistry differs widely from its other meanings in common usage. Pupils frequently find it difficult to distinguish these usages at this stage, and the Study sheet aims to help them over this.

The halogens. Besides describing many of the beneficial ways we use these elements, the Study sheet questions the responsibility of the scientist who developed chlorine as a weapon in the First World War, and it points out the harmful effects of bromine in petrol. (The idea for the first part of this Study sheet is drawn from Holmyard, E. J. (1925) *An elementary chemistry*. Edward Arnold.)

Chemicals and rocks aims to give pupils an idea of the great variety of rocks and minerals, and to show how chemistry can be used to study them.

Useful books on the same theme as the above Study sheet include:
Schools Council Integrated Science Project (1973). *Rocks and minerals*. Longman/ Penguin Books.
Brennand, S. (1973) *Gold and granite*. Puffin Books.
Allan, M. (1972) *What do we know about the Earth*. Blackie.

B10.1
How can we get metals from their oxides?

The relative reactivity of metals is assessed from previous experiments and a reactivity table drawn up. The pupils do experiments to find whether carbon and hydrogen reduce certain metal oxides. They find that the residue left when rust is heated (see B9.4) is an oxide of iron.

A suggested approach

Objectives for pupils
1. Recognition that there is a pattern in the vigour of reactions

The first part of this section concerns the building up of a reactivity table from evidence already discovered in previous experiments, the most important of these being the burning of metals in oxygen, B4.1 and B8.2, and the measurement of electrode potentials in B6.5. The lesson should start therefore with a discussion of the results of these experiments. From these results a reactivity series may be built up,

2. Awareness that an order of reactivity enables reactions to be predicted
3. Knowledge of the order of reactivity of some common metals
4. Understanding of the meaning of the words reduce* and reactivity

use being made both of the electrode potentials and of general reactivity with oxygen. Having done this, suggest that we need something 'more grasping' than the metal if we want to remove oxygen from it and get it back to its original form. 'If one metal can be seen to react more vigorously than another metal, will the first metal be capable of removing oxygen from the oxide of the second?' Once again it must be stressed that 'vigour of reaction' is by no means an infallible guide as so many factors affect it. But it does give one a rough guide to the reactivity series and it is easily visualized by the pupils. They could predict, for instance, that magnesium would remove oxygen from copper oxide. But metals are relatively expensive and you can explain to the pupils that hydrogen and carbon are two frequently used non-metals. 'Where do they come in the reactivity series?' The answer can be found by heating metal oxides with carbon or in a stream of hydrogen. To make sure that the metal oxides are being reduced, attempts should be made to detect carbon dioxide and water respectively in the products. The experiments are described below.

Experiment B10.1a

Will carbon turn copper oxide to copper?

Apparatus

Each pupil (or pair) will need:

2 hard-glass test-tubes, 100 × 16 mm

Test-tube holder

Bunsen burner and asbestos square

Teat pipette

Spatula

Limewater

Copper(II) oxide

Carbon (dry powdered wood charcoal)

Procedure

There is no *Experiment sheet* for this short experiment; instructions should be given by the teacher as follows.

Place a spatula measure of copper(II) oxide and a spatula measure of carbon (powdered wood charcoal is suitable) in a test-tube and mix them well by shaking. Heat the test-tube for some time in a near vertical position to ensure that any carbon dioxide formed does not 'fall out'. Then allow the test-tube to cool, withdraw a sample of the gas above the heated solid by means of a teat pipette, and bubble through a *small* quantity of limewater in another test-tube.

Carbon dioxide will be detected in this case.

Experiment B10.1b

Will hydrogen turn copper oxide to copper?

This experiment **must** be performed by the teacher.

Apparatus

The teacher will need:

Plastic safety screen

Hard-glass test-tube, 125 × 16 mm, with small hole blown near closed end, bung, and short delivery tube

Asbestos paper strip

Stand and clamp

Bunsen burner and asbestos square

(Continued)

Procedure

Place a safety screen between the apparatus and the class. After placing about two spatula measures of oxide in the test-tube as illustrated, pass a gentle stream of dry hydrogen through the tube. Wait until all the air has been displaced, then light it at the hole and turn the gas down so that the flame is only about 2 cm high. The tube requires only gentle warming for the reduction to take place.

The copper oxide will turn pink, showing that it has lost its oxygen; and if the heating has been moderate enough, water will be seen con-

*Some caution is needed here; see the *Handbook for teachers*, Chapter 2, pages 31–34.

Taper

Rubber tubing (to connect delivery tube to gas tap)

6 V bulb and holder

6 V battery or alternative d.c. supply

Connecting wire

Hydrogen cylinder

Copper(II) oxide

Magnesium oxide

Zinc oxide

Lead(II) oxide

Residue from heating rust (Experiment B9.4)

densed at the closed end of the test-tube. The pink residue, which consists of a sintered powder, can be shown to be a metal by touching it with the two leads from a bulb and battery.

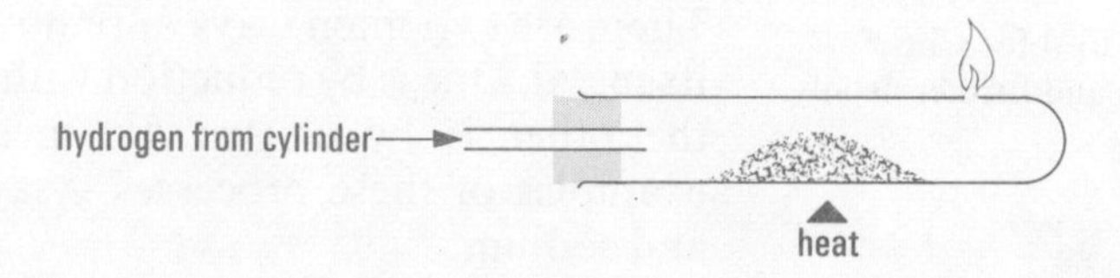

The above experiment may be repeated using magnesium oxide, zinc oxide, and lead oxide and the residue from the heating of rust. Note that in the case of lead oxide the combustion tube will be attacked and damaged. This can be avoided by putting the oxide on a piece of dry asbestos paper inside the tube. Only the lead oxide and the rust residue will be found to be reduced. Reducing the 'rust residue' to iron will show that it is iron oxide.

Suggestions for homework

1. Read the Study sheet *Chemicals and rocks.*
2. From your experience so far, what would you expect to happen when hydrogen is passed over the following metal oxides, heated as necessary: (*a*) mercury oxide, (*b*) calcium oxide, and (*c*) nickel oxide?
Give reasons for your predictions.

New words

Reduce
Reactivity

Summary

Pupils should now be aware that reactive elements are likely to remove the oxygen from oxides of less reactive elements when they are heated together. In particular they should know that carbon will reduce copper oxide, and that hydrogen will reduce iron, copper, and lead oxides. They should understand the meaning of the words reduction and reactivity.

B10.2
How metals are obtained from their ores

In this section some aspects of the industrial production of metals are discussed. It is an opportunity for pupils to become aware of the activities of the chemical industry, and to appreciate the usefulness of chemistry to Man.

A suggested approach

Objectives for pupils

1. Awareness of the usefulness of chemistry
2. Knowledge of the industrial extraction of some metals

The removal of oxygen from oxides has numerous practical applications. You may tell the pupils that the production of metals from naturally occurring metal ores (mostly sulphides and oxides) has been carried out since the dawn of civilization and long before the chemical processes involved were understood. Iron, for instance, occurs as the oxides haematite or magnetite. How could iron be obtained from this ore? By removing oxygen with carbon? In what form is carbon obtainable? Wood charcoal – yes, that was used but

Useful films

'Making iron'

'Study in steel'

See Appendix 4 for a brief description and further details

supplies proved to be inadequate for demand. Coal contains too much tar and other impurities. Coke proved to be the answer. Now metallurgists are trying to find ways of using oil in blast furnaces.

There are two main ways of removing oxygen from an ore to obtain its metal. One is by reduction with coke or some other chemical and the other is by electrolytic means. You might discuss several examples of these processes – say lead, iron, copper, aluminium, and sodium.

At this point a film of one of these processes may be shown. There are a number of good films available, among them are 'Making iron' and 'Study in steel'.

Suggestions for homework

1. Write an account of the extraction of one metal from its ore and comment on the usefulness of the metal in everyday life.
2. Find out about the ores of the more common metals. What is the chemical composition of these ores? Can you see any pattern linking the type of ore with the position of the metal in the reactivity series?

Summary

After this section pupils should know something of the principles of the extraction of metals from their ores.

B10.3 Can metals be displaced from solutions of their salts?

The concept of competition among the elements is illustrated by the displacement of metals from solutions of their salts by metals higher up the reactivity series.

A suggested approach

Objectives for pupils

1. Knowledge of the reaction between metals and metal salt solutions
2. Ability to use the idea of an order of reactivity

Start this section with a discussion of the reactivity series, reminding the class of the experiments they performed in section B10.1, and again of the Experiment B6.5a and b. The purpose of the experiments in this section is to show that the position of a metal in the reactivity series can be used to predict its behaviour in displacing other metals from their salts. A penknife blade dipped in copper sulphate solution becomes covered with copper, showing that iron displaces the less reactive copper from a solution of its salts. If it is a general rule that the more reactive metals displace the less reactive from their salts, it should be possible to predict a number of reactions. Let the pupils make predictions and then test them as described in the experiment below.

Experiment B10.3

More competition among the elements

Apparatus

Each pupil (or pair) will need:

Experiment sheet 39

6 test-tubes, 100 × 16 mm, in test-tube rack *(Continued)*

Procedure

The pupils should take samples of the metals copper, iron, lead, magnesium, tin, and zinc and place a small piece of each metal in a little copper nitrate solution, to see if the metal replaces the copper in the solution. (Ensure that the metal samples are free from grease, and that they are not handled more than is necessary.) If it does, a

Samples of the following metals: a suitable form for each metal is indicated, but this choice is not essential:

Copper – 1 cm square pieces of foil

Iron – 3 cm iron wire or nail

Lead – 1 cm square pieces of foil

Magnesium – 3 cm lengths of ribbon

Tin – 1 cm square pieces of foil

Zinc – 1 cm square pieces of foil

Approximately M solutions of:

Copper(II) nitrate

Lead(II) nitrate

Tin(II) chloride

Iron(II) sulphate

Magnesium sulphate

Zinc sulphate

copper coating will be seen on the surface of the metal, but several minutes may be needed before the coating is clearly visible. If no action appears to have taken place, pupils should remove the sample from the solution and gently scrape the surface to see if there is any deposit.

The experiment should then be repeated but this time placing a sample of each metal in iron sulphate solution; then lead nitrate solution, magnesium sulphate solution, tin chloride solution, and zinc sulphate solution may be used.

The pupils should then use their results to draw up a list of the metals such that any metal will displace all those below it in the list from solutions of their salts.

The procedure is described in *Experiment sheet* 39 reproduced below.

The observed displacement series for this list of metals is: magnesium, zinc, iron, tin, lead, copper.

Experiment sheet 39

Using samples of the metals copper, lead, zinc, magnesium, iron, tin, and solutions of their salts (say their chlorides or nitrates), perform experiments to see which metals will displace others from solution.

1. Start with a solution of copper salt. Place some in a test-tube and put a bright iron nail in it. Leave it for a few minutes. What do you see?

Repeat the experiment using, in turn, other metals instead of iron. Draw up a table, on a sheet of paper inserted opposite, to make notes of all that you observe.

2. Now do the whole experiment again using a solution of lead salt instead of the copper salt. Add the observations to your table.

3. Continue by doing the experiment with solutions of other metal salts. Using the results, make a list of the metals in an order such that any metal will displace all those below it in the list from solutions of their salts.

Suggestions for homework

1. Read the Study sheet *Competitions*.
2. Compare the merits of lead and zinc for use in outdoor water pipes.
3. Supply the class with a further displacement series and ask the pupils to predict and describe what should happen when certain metals, which they have not already used in their practical work, are added to various salt solutions.

Summary

Pupils now know that reactive metals displace those that are less reactive from solutions of their salts. They have seen that a reactivity series can be built up for these metals, and that such a series can be used to predict the outcome of displacement reactions.

B10.4
Can non-metals be displaced from solutions?

In a similar way to that shown in B10.3 non-metals are seen to be able to displace less reactive non-metals from their salts.

A suggested approach

Objectives for pupils

1. Knowledge of some non-metals, and the reaction between them and solutions of their potassium salts
2. Ability to use the idea of an order of reactivity

The lesson on this subject may begin with a discussion of the homework set after the last section and lead to the question 'Can the same sort of displacement series be built up with the non-metals as with the metals?' This is a more difficult problem, but the pupils should be allowed to try to find the answer.

Begin by reminding them of the appearance of sulphur and the halogens, so that they know what to look for in their experiments.

Experiment B10.4a

Apparatus

The teacher will need:

Filter flask, fitted with delivery tube, bung, and tap funnel, with which to generate chlorine

4 stoppered bottles, about 50 cm^3 size

3 test-tubes, 150 × 25 mm, in rack

Concentrated hydrochloric acid

Potassium permanganate

Bromine

Iodine

Sulphur

Distilled water

The appearance of sulphur and the halogens

This experiment must be carried out by the teacher, preferably in a fume cupboard.

Procedure

Begin by filling a large test-tube with chlorine so that the pupils can see that it is a green gas. Do this by dropping two or three drops of concentrated hydrochloric acid from a tap funnel onto well-ground crystals of potassium permanganate contained in a filter flask (see diagram). Take care not to inhale the gas, and do not let the acid drop too rapidly onto the permanganate. Pour out a little bromine (a red liquid – **care** – very corrosive, wear protective gloves) into a test-tube standing in a rack, to avoid spilling on the hands. Place a few crystals of iodine into a third test-tube. Iodine crystals will be seen to be black, but produce a deep purple vapour when gently heated.

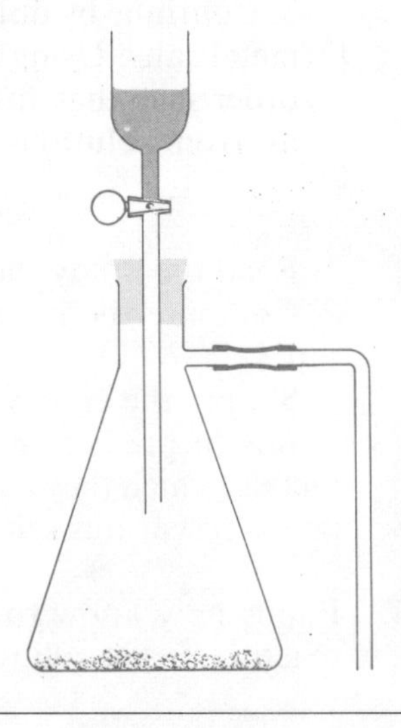

Put 5 cm^3 of water in each of the three bottles. Bubble some chlorine into the first one, place a few drops of bromine into the second, and a few crystals of iodine into the third. Stopper each bottle securely and shake the contents. Show the pupils the colours of the solutions that are formed (green, red, and brown respectively). Show the pupils a sample of sulphur, and by shaking a little with water in the fourth bottle, show them that it is insoluble.

After showing the pupils sulphur and the halogens in this way, let them do Experiment B10.4b (*Experiment sheet* 40). In this experiment they can try bubbling oxygen into solutions of a sulphide, a chloride, a bromide, and an iodide, for a few seconds. Nothing happens. They should then try the effect of chlorine (as chlorine water); this will give some positive reactions. Bromine will be displaced from a bromide by chlorine. They may use this bromine water to displace iodine from an iodide solution. Details of experiments which they may try are given below.

Experiment B10.4b

Competition among some non-metallic elements

Apparatus

Each pupil (or pair) will need:

Experiment sheet 40

4 test-tubes, 100 × 16 mm, and test-tube rack

Polythene bag, cork or bung, and 5 cm length of glass tubing

Delivery tube – glass tube from teat pipette

Solutions of potassium sulphide, potassium chloride, potassium bromide, and potassium iodide

Chlorine water

The teacher will need:

Cylinder of oxygen

Procedure

Each pupil should collect half a test-tube of each of the following 1M solutions and place them in a test-tube rack:

Potassium sulphide
Potassium chloride
Potassium bromide
Potassium iodide

The pupils should then bubble a little oxygen through each to find out if any change takes place. This may be done directly from a cylinder of the gas. With a large class it may be convenient for the teacher to inflate a small polythene bag from an oxygen cylinder for each pupil, who can then squeeze out the oxygen and bubble it through each solution.

The experiment can then be repeated using the same solutions but with chlorine in place of oxygen. Owing to the poisonous nature of chlorine and the difficulty of providing adequate fume cupboard facilities, chlorine water should be provided. Each pupil should add *a little* chlorine water to the solutions. Displacement of bromine and iodine will be observed.

The displaced bromine may itself be used to displace iodine.

Experiment sheet 40 is reproduced below.

Experiment sheet 40

Having done some experiments on displacement of metals from solutions of their salts you will now investigate whether non-metals can be displaced in a similar way.

Fill four test-tubes to a depth of about 2 cm with solutions of potassium sulphide, potassium chloride, potassium bromide, and potassium iodide. Bubble some oxygen gas through each tube. In which solution do you notice any change?

Experiment sheet 40 *(continued)*

Empty and wash the four test-tubes and re-fill them as before. Add about 1 cm^3 of 'chlorine water' to each tube ('chlorine water' is a solution of chlorine in water. **Do not hold it near your nose or eyes.** What happens this time to each tube?

Try adding a little of the contents of the tube which contained potassium bromide solution to a fresh portion of potassium iodide solution. What happens?

Try to explain all your observations.

A discussion on the results of these series of experiments will lead to the establishment of an order of reactivity for sulphur and the halogens. Some time may now be spent in building up a more complete reactivity series.

Suggestion for homework

Draw up a list of the different ways that you have been able to build up reactivity series for metals and non-metals. Make a comparison of the various lists.

Summary

Pupils now know that chlorine displaces bromine and iodine from solutions of potassium bromide and iodide, and that bromine displaces iodine from potassium iodide solution. They should be aware of the reactivity series sulphur, chlorine, bromine, and iodine, and be able to use it to predict the likely outcome of chemical reactions involving these elements.

B10.5
The family of halogens and their industrial production

The family properties of the halogens as a group of elements are discussed and their industrial production and general uses are demonstrated in the film loops.

A suggested approach

Objectives for pupils

1. Awareness of the idea of families of elements
2. Knowledge of the methods of production and of the main uses of the halogens

The teacher will need

Film loop projector

Film loops

1–11 'Fluorine manufacture' (with notes)

For the first time a group of elements, within the main groups of metals and non-metals, has been introduced. It may be pointed out to the pupils that the elements chlorine, bromine, and iodine, which they studied by displacement in the previous section, are all closely connected in chemical properties and that they do in fact represent a chemical family. Another member of the family is the element fluorine which is too dangerous to be demonstrated. A brief character sketch may be given of each element and its production and uses may be discussed. A series of film loops has been made for this purpose. They show the industrial preparation and main use of each of the halogens, and they indicate the depth of treatment which is intended at this stage.

1–12 'Fluorine compounds uses' (with notes)

1–13 'Chlorine manufacture' (with notes)

1–14 'Chlorine uses' (with notes)

1–15 'Bromine manufacture' (with notes)

1–16 'Bromine uses' (with notes)

1–17 'Iodine manufacture' (with notes)

1–18 'Iodine uses' (with notes)

Useful film

'Treasure trove'

See Appendix 4 for a brief description and further details

Suggestions for homework

1. Read the Study sheet *The halogens*.
2. Choose one of the halogens and find as many uses for it and its compounds as you can.
3. Make a summary of the work you have done in this Topic.

Summary

Pupils have seen something of the methods of production and the uses of the halogens.

Part 3 Appendices

Appendix 1

School and class organization

Introduction

A new teaching scheme is bound to create a great many practical problems for those who use it. What will it cost in money and man-power? How much time will be needed? How will the balance of the curriculum be affected? What about examinations? Questions like these are among the tests by means of which practising teachers must always decide how to marry educational ideals with the realities of the classroom and laboratory. They deserve tangible and practical answers.

Although the questions that teachers ask are likely to be very similar in one school and another, it does not, unfortunately, follow that there is a unique set of solutions. Circumstances vary. Sometimes there may be ample time but not enough laboratory space. Elsewhere, shortage of laboratory technicians, or of money, may be the deciding factors. At other schools, again, the ease with which teachers can introduce new schemes of work may be influenced by more general considerations – a tradition of general science in the first few years, for example, or by policy on the timing of O- and A-level examinations. This diversity is no surprise, but a familiar and a valuable ingredient in the educational system. However, one inescapable consequence is that it must be for individual teachers themselves to decide how to make use of the Nuffield Chemistry scheme in their own schools. What follows, then, is not a general recipe for solving once and for all the problems of organization that will be created, but a compendium of some of the raw materials with which teachers may wish to work out their own solutions.

Time allocation

The alternative schemes have been devised on the assumption that the time available for teaching chemistry amounts to two periods (of 40 to 45 minutes) each week in the first two years, and three periods a week in the last three years of a five-year course:

Years 1 and 2 (Stage I)	1 double period in a laboratory *plus* 1 homework period
Years 3, 4, and 5 (Stages II and III)	1 double and 1 single period in a laboratory *plus* 1 homework period

This allocation of time is that recommended in 1961 by the Association of Science Education. Its adoption by the Nuffield Science Teaching Project is based on the belief that individual practical work by pupils is essential to a proper treatment of chemistry in schools. One double laboratory period each week is necessary if the investigations of even the youngest pupils are to be fruitful. After the two introductory years, when the investigations are more complex, and when explanations in terms of conceptual models are more prominent, extra time – an extra period – is needed for class discussion and the planning of investigations.

Not all schools wishing to try Nuffield Chemistry will have this allocation of time. Some may indeed have more but, unfortunately, some may have less than is recommended. It would be folly to pretend that circumstances are otherwise, even though the allocation suggested is modest compared with that usually made for other core subjects in the curriculum. The scheme will often have to be adapted to suit the time allotted to chemistry, and the way in which this is done must be left to the judgment of the teachers concerned. In seeking adaptations, however, one salient consideration should not be forgotten. Access to a laboratory is crucial for almost every lesson in the scheme. Without it, an investigational approach is almost impossible.

Costs

There cannot, of course, be a precise figure for the cost of introducing Nuffield Chemistry. Much will depend on how well equipped a school may be to start with. As with other schemes for teaching chemistry at this level, the chief needs are chemicals and not very complicated glassware.

Certain items of capital equipment are, however, desirable to make the best use of laboratory time. Of those for Stage I, a direct-reading balance, a centrifuge, cylinders of hydrogen, oxygen (and nitrogen if Alternative B is adopted), and two gas syringes are the most expensive. If the film loops are used, both they and a suitable projector will be required. None of these items is indispensable, but their absence will make additional demands on the teacher.

The range of chemicals and apparatus does not differ markedly from that already used in schools, although there are a few chemicals (such as lead bromide) which are not commonly used at this level, and a few pieces of apparatus (such as a special spatula) which will be found helpful.

Even though some additional expenditure may be incurred initially, experience in schools has shown that the annual expenditure thereafter should not be noticeably increased – provided that practical work by pupils already played a major part in the teaching.

Laboratory organization

Personal investigation at the laboratory bench plays an integral and essential part in the Nuffield Scheme. There is more practical work for pupils than has been customary in the past, and much of it is decidedly novel in character. There is a place for experiments performed by the teacher, or by the teacher and a few pupils, but activity of this kind should be infrequent. The constant danger is that pupils may be robbed of the sense of discovery, and that they may encounter too many important matters at second-hand. As teachers, we should constantly ask, 'Why is this experiment being done by me and not by the class?' Only the most convincing answers should be acceptable.

These principles create problems for the teacher. Adequate and efficient assistance in the laboratory is an invaluable step towards their solution. But so, too, is careful organization which can do much to remedy unsuitable and inadequate conditions.

Time is an important factor. A great deal of class investigation is possible in the time allocation recommended for Stage I, but much can be done even where this is not possible. A single period should not be regarded as too short for individual work. Many worthwhile and rewarding experiments can be done in twenty minutes, or even ten, and it is often better to break down a fairly lengthy piece of work into two or more parts, particularly when deductions from one part are used in the next. It is clear, however, that such short spells of practical work are possible only if the time spent in issuing apparatus and materials, and tidying up afterwards, is reduced to a minimum. From the experience gained in schools using the scheme, some simple ideas and devices have emerged. These are outlined below in the hope that they may be of use to others. Few of them are original and many teachers will already know of them.

Order and discipline

The kind of laboratory work described in this book has been found to arouse considerable interest and enthusiasm among pupils. They become so keen on their allotted tasks that their chatter can rise to noisy heights and they tend to move around excitedly so that disorder results. This inhibits effective work. For an investigation to be carried out satisfactorily, without unnecessary accidents, there must be a calm and orderly atmosphere in the laboratory. The children can then focus their attention on what they are doing and begin to think about their observations and results. We want to encourage the occasional 'bright idea' and must create conditions in which this can emerge.

The teacher has a positive part to play. The objective should be an atmosphere of friendly collaboration between class and teacher. With good preparation, and if the class is carrying out its work fairly quietly (some talking must be allowed), the teacher need only watch that all is going well, moving from group to group and questioning them about what they are doing. When an instruction needs to be given to the whole class, silence should be established by waiting until pupils can safely remove their attention from the experiment. Experienced teachers know that attempts to shout above the noise of children doing interesting practical work are fatal to good order. A film, 'Exploring chemistry', which includes typical Nuffield lessons, records a successful attempt to capture this spirit. See Appendix 4 for further details.

Distribution of apparatus

There are two ways of doing this efficiently. One is to group the various pieces of apparatus item by item on a side bench before the lesson begins. Pupils then collect what they need and afterwards return it to its original place. It is then easy to check that glassware is clean when returned, and to check breakages. Convenient methods of storing individual items in class sets so that they may be moved from and to permanent storage places are described below.

An alternative method is to store frequently used apparatus in cupboards and drawers at the pupils' working places. This minimizes movement about the laboratory. It also eliminates time being spent by the teacher or technician in putting out apparatus before the class assembles. With frequent checks and replacement of missing items (which can be done in the odd few minutes that are sometimes to spare at the end of a lesson), this method can be quite effective. The main disadvantage, apart from the danger of pilfering, is that dirty glassware tends to accumulate unnoticed. Pupils should be encouraged always to bring apparatus that they cannot clean to the teacher, who can either give assistance with removing stains (much can be done with a little concentrated hydrochloric or nitric acid while the class is cleaning up), or issue a clean item from a reserve store, leaving the cleaning to be done later by the technician or himself.

It is useless to expect pupils to take care over cleaning apparatus and laboratory benches unless they are provided with adequate means of doing so. Both glass cloths and bench cloths should be available, and their different uses emphasized at frequent intervals. Bench cloths of the plastic-sponge type are most suitable as they are absorbent, unlike many old-fashioned bench swabs; they should be rinsed well after every use, and kept near the sinks. Glass cloths, in class sets, are best hung over wooden rails fixed to a side wall or under a side bench. If stored in drawers they are rarely dry when needed. Pupils should be given some instructions about the best methods of cleaning glassware. A supply of hot water is useful for this and a few 'squeeze' bottles of detergent should be available in the laboratory.

Storing apparatus

Easy transport of class sets of apparatus from permanent storage places to the laboratory saves much time and breakage. Some useful devices are now described; others will easily come to mind.

1. A rack for conical flasks made to the same design as the old-fashioned egg rack, that is, a wooden baseboard with corner uprights supporting a sheet of plywood in which circular holes have been drilled. The flasks are stored neck downwards so that they can drain. Dirty flasks are easily detected. The rack should hold a class set of flasks with a few spares.

2. Fruiterer's 'tomato trays' can be bought very cheaply. They are of standard size and stack vertically. Beakers are conveniently stored in them as well as apparatus for the teacher (one experiment per tray). They are also useful for putting aside half-finished experiments. Each working pair of pupils should be identified by a number and the trays should be divided into numbered squares with chalk.

3. If burners are not permanently attached to gas points at working benches, they can be stored in a wall rack. Slots about 20 mm wide and 25 mm deep are cut into a plywood strip 75 mm wide. This is fastened to a wall (or the side of a cupboard) using wooden

battens 25 mm thick to keep it away from the vertical surface. The bases of the burners slide down between wall and plywood strip with the chimneys in the slots.

4. Many items of apparatus can be stored conveniently in stout cardboard boxes; suitable sizes (all figures in inches) are $8 \times 5 \times 2$, $10 \times 8 \times 4$, and $10 \times 8 \times 8$. The smallest sized box can be used for items such as spatulas, microscope slides, asbestos paper strip, and other small items. The medium-size box is convenient for the storage of corks, evaporating basins, teat pipettes, and so on. The $10 \times 8 \times 8$ inch size is large enough to contain many of the sets of apparatus for class use, for example distillation kits for the distillation of crude oil. The boxes are best stored on adjustable shelving so spaced that only two or three of the medium-size boxes can be stored on top of one another. This avoids the necessity of moving a large pile of boxes to obtain the bottom one, and there is less danger of the boxes falling over.

Issuing chemicals

One plan is to put out the chemicals needed in small piles on pieces of paper, in watch-glasses, or in shallow dishes, with a spatula beside each. Pupils then help themselves. Identification is made easy if the bottles from which the separate substances are taken stand alongside. The quantity put out should be rather less than is likely to be needed, as unused material should not be returned to a stock bottle.

Chemicals needed frequently can be stored in small labelled containers, such as corked sample tubes, in a rack of similar pattern to a test-tube rack. Liquid reagents in common use can also be kept in small bottles in racks; the racks can then easily be transported to the pupils' working space when needed. Other liquid reagents can be dispensed as needed in small conical flasks. It is important that these are clearly labelled.

Smaller amounts of solid reagents than have been customary are recommended in the Nuffield proposals. Pupils need some guidance about how much to take. Various 'units' such as 'the size of a pea', 'the size of a rice-grain', and a 'nibful' have been used in the past, but are not really very helpful. A special spatula has been designed to solve this problem. One end of this is grooved to serve as a convenient measure which holds about a 'nibful'. In the publications it is called 'a measure'.

Disposal of solid waste is frequently a problem. Receptacles near each sink are the most effective solution, removing the temptation to use the sink as a wastebox. Stoneware jars or plastic bowls are suitable.

Size and type of apparatus

Much time can be saved for both class and teacher by using the smallest sizes of apparatus suited to the purposes of the experiments and the manipulative skill of the pupils. Thus 100×16 mm test-tubes, 100 cm^3 beakers, and 100 cm^3 conical flasks are adequate for most practical work. Small dishes (or watch-glasses) save time in evaporation. Very substantial economies in chemicals are possible by working on such a reduced scale. Time is also saved in preparing solutions and replenishing reagent bottles.

Cheap glassware, usually made of 'soda glass', should be avoided, except for reagent bottles and glass tubing. Apparatus made from borosilicate glass is more expensive to buy, but its longer life more than compensates for this, and it is much safer to use.

Use of bunsen burners
The bench should be protected by standing the burner and the apparatus on an asbestos tile. Wooden splints are convenient for lighting burners from a pilot light on the teacher's bench. Wax tapers are costly and also lead to trouble with blocked jets on Bunsen burners. Pieces of paper, even when folded into spills, are untidy and can flare up dangerously.

Stoppers
Corks have a limited use especially where gas-tight apparatus is essential. Rubber bungs are superior for such purposes, but if corks must be used their porosity is greatly reduced by dipping them into *hot* molten paraffin wax for a few seconds. A power drill is useful for boring both corks and rubber bungs provided that a special clamp is used to hold them. (It is dangerous to hold them in the hand during this operation.) If cork borers are used, they must be kept sharp. The conical sharpeners sold for this purpose are rarely needed. A few gentle strokes with a medium carborundum stone around the outside edge should suffice.

Corks and rubber bungs, especially those carrying thistle funnels and delivery tubes, should be stored with the pieces of apparatus with which they are to be used. Much time in finding and boring suitable replacements can thus be saved. If glass tubing (or a thermometer) has to be removed from corks or rubber bungs, a slightly over-sized cork borer, lubricated with glycerol, should be put over the tube and into the cork. Borer and cork are then rotated in opposite directions until the tube is freed.

Pieces of apparatus with ground glass cones and sockets have many advantages. Complex assemblies can be built up rapidly from standard components. Such apparatus is rather fragile for young pupils, and is still relatively expensive. Yet many schools use it in sixth forms, and there seems no reason why fourth- or fifth-year pupils should not become accustomed to it. For experiments performed by the teacher, it has everything to commend it.

Plastic bottles
Empty household detergent containers make excellent wash bottles. The plug is removed and replaced by a rubber bung fitted with a bent tube drawn out to a jet at one end. Smaller plastic bottles can be purchased for a few pence each at multiple stores. They are useful for reagents and indicators, especially if fitted with a rubber stopper and teat pipette. Another use is as calorimeters for measuring heats of reaction.

Thermometers
The conventional thermometer, usually about 30 cm long, has a high casualty rate in the hands of schoolchildren. The much shorter variety, less than half as long, with a thick glass reinforcement at the bottom of the bulb, will survive almost any treatment short of being dropped on to the floor. They are certainly robust enough to be used as stirrers – a profitable use since it leads to the detection of many unsuspected energy transfers.

Filtering
This can be a time-consuming process. Three methods of speeding it up are recommended.
1. By using 'rapid' filter papers in the conventional filter funnel. These are coarse and tough, are rather expensive, but can be used several times.
2. By 'force filtering'. Young pupils can handle small filter flasks and small Buchner or Hirsch funnels quite efficiently. Filter pumps, or a low pressure line, are also needed.
3. By centrifuging. This has many advantages. In quantitative work a precipitate can be obtained, separated, washed, dried, and weighed in the same test-tube.

Safety precautions

Pupils should be warned about potential dangers in the laboratory. These are mostly obvious to the teacher, who will be alive to the hazards of liquids being forcibly ejected from containers on heating, burners which have 'lit back', hot tripod stands, cuts from glass tubing used unwisely, explosive gaseous mixtures, and the like. A knowledge of elementary first aid is essential to every chemistry teacher. A short list of safety rules, *Safeguards for school laboratories*, is obtainable from the Publications Officer, the Association of Science Education, and should be displayed in all school laboratories. The A.S.E. also publish *Safety in the lab.* (sold in packs of 50 or 100) for distribution to each child. The Department of Education and Science publish *Safety in science laboratories*, No. 2 (which supersedes *Safety at schools*, E.P. No. 53) for teachers and laboratory assistants. This should be available in every science department. Teachers may also like to consult Everett, K. and Jenkins, E. W. (1973) *A safety handbook for science teachers*, John Murray.

Experiments which entail special hazards, such as those involving substantial quantities of toxic or corrosive materials, inflammable gases, or vigorous reactions, will be done by the teacher. Wherever there is the slightest risk of an explosion or the ejection of corrosive substances from containers, a transparent safety screen must be placed between apparatus and pupils. The teacher should also protect himself by this means whenever possible. Protective eyeguards and rubber or plastic gloves are also recommended, both for teachers and pupils, where these are necessary. A chemistry laboratory should be regarded as a place in which potentially hazardous operations can be performed with safety. This is a part of general education which is by no means insignificant.

Dealing with noxious gases

'This experiment should be carried out in a fume cupboard or under a hood' is unhelpful advice to the teacher whose laboratory has no hood and only one tiny fume cupboard. Since all noxious gases encountered in an elementary chemistry course are acidic (except ammonia which is not really much of a nuisance), they can be absorbed safely by solid alkaline reagents such as sodium hydroxide or soda lime (coarse grain, for example 6–10 mesh). This has the advantage of eliminating possible 'suck back'. Arrangements are made to lead excess gas, from any apparatus in which it is used, through a glass tube into a flask containing the absorbent. The tube should extend to just above the surface of the soda lime or sodium hydroxide, and the neck of the absorption vessel should be packed with cotton wool. It is quite safe to remove excess chlorine, for example, in this way. If sodium hydroxide is used, the flake variety is preferable because it is cheap and because a larger surface is thereby exposed. The use of an absorbent provides an additional teaching point – it gets very hot and is thus a good example of an exothermic reaction.

Appendix 2

Pupils' laboratory records

The purpose of records

Writing up practical work plays a predominant part in present science teaching. Indeed it is not unknown for careful 'writing up' to be done without any experiment having been performed! On the one hand, written work can be a dreary chore for the pupil and an almost fruitless burden for the teacher who has to mark it. On the other hand, behind it all is the conviction that keeping a record is 'essential training' and a 'good thing'.

We could resolve this dilemma if we clearly and boldly faced two questions: What is the purpose of keeping a record? Does the pertinence of this exercise change as a pupil gets older?

We shall tend to answer these questions in differing ways but the Nuffield proposals are based on consideration of the following points:

1. The chief reason for making a pupil put something on to paper is to sharpen his understanding. What must be avoided is turning it into a chore, and only the teacher can tell whether or not the written work is of value.
2. Building up a notebook is in most cases enjoyable and stimulating if it is *personal*; this can rarely be so if the material is dictated and made uniform throughout the class.
3. Much of the character of written work as done at present is formed by the need to produce, and file for future reference, answers to conventional examination questions. This often persists throughout the school course from junior to senior forms.
4. The attitudes of pupils to the value of what they do are known to change considerably as they grow up. Furthermore, we can adjust the demands made throughout the school course so that, by the time pupils are sixteen, we have brought them to a level of attainment appropriate to their ability and interest.

How to keep record books and how to mark them

In the early part of the course, much that is done does not need permanent record. This not only applies to details of practical techniques, such as filtering and evaporating, but also to class discussions, and to many quick exploratory experiments. Young pupils want to be doing most of the time and must not be hampered by too much written work, which they often find difficult, and at which most of them are very slow. They should be encouraged to regard their notes as diaries in which they record, briefly and simply, the more important things that they do, and the significant speculations and conclusions that lead to and emerge from their experimentation. A good deal of guidance from the teacher will be essential at the beginning of the course. Methods of guidance – half-completed tables or sentences on the blackboard, skeleton notes, help in drawing labelled diagrams – are well known. However, as with all diaries, the individuality of the compiler will soon become apparent and some pupils will want to include samples of substances that they have made (sometimes at home), cuttings and illustrations from newspapers and periodicals, chromatograms, indicator colours, charts, and a host of other items. This should be encouraged but not insisted on. It is one way in which enthusiasm and interest will show themselves; these cannot be imposed but should be fostered.

It is, however, important to realize that this procedure alone is not enough, if, at the end of a term or when a test or examination is held, the pupil is insufficiently sure of what he has achieved.

Although the path followed in an exploration may not have been straight, the goal, once reached, should be clear and definite. The pupils need to 'know what they know' and not be left in any uncertainty about what conclusions have been reached at the end of a piece of work. It has been found that one of the best ways of doing this is to get the pupils to write their own summaries, probably at the end of each Topic. These should be on separate sheets of file paper, and, when collected together at the end of each year or stage, would form a clear account of the work done and be useful material for use in revision.

The record books will need to be looked at, and sometimes corrected, by the teacher; this could be done in batches and the work in them discussed with the pupils at convenient times during lessons. The pupil's summaries should be carefully corrected. Recurrent errors should be made the subjects of class discussions. There is value also in letting pupils have the opportunity of seeing each others' records and discussing these among themselves.

Training in recording observations and speculations on the lines indicated above will mean that:
1. The pupils regard the immediate setting down of observations and results as essential. The printed sheets will help here.
2. They maintain a reasonable standard of neatness and accuracy.
3. They regard completion of observations and measurements as a halfway stage in an experiment because, once they have them, they still have to find out what they mean.

Towards the end of the course we can say that the pupil's laboratory notebook should be a sort of journal of the progress of the work. It should contain more than a mere account of what operations he carried out and what results followed: when completed it should be a valuable account of the term's activity and include a summary of any discussion that led up to some practical work, the reasons for doing an experiment, the planning of the experiment in terms of apparatus, chemicals, and procedure, in addition to a clear statement of the observations made, a discussion of their significance, and a note of any suggestions for further work that emerged. These should lead on naturally to the account of the next piece of work. Thus the notebook has at least two uses:

1. At the time of writing, having to think out in clear English what to put down helps to clarify the topic for the pupil.
2. It serves as a useful reference and means of revision, often easier for the pupil than printed books.

The form and purpose of *Experiment sheets I*

To help the pupil in this quite difficult job of making a good record, we suggest that the notes be written on sheets of file paper interleaved with the printed sheets of instructions for the laboratory work called *Experiment sheets*. These have been written with this arrangement in mind and frequently pose questions which the pupil is expected to answer in the spaces left on the sheets.

Appendix 3

Experiments for Stage I

This appendix contains all the experiments described in this book; further experiments can be found in *Collected experiments.* The first two lists detail the experiments in the Alternative A and Alternative B schemes respectively. The third list (page 240) contains the sheet numbers and titles in the pupils' book *Experiment sheets I* and their correlation with the *Teachers' guide* experiments.

Experiment sheets I contains laboratory sheets for experiments to be carried out by pupils, bound together in one volume. This has been done because it is more economical than producing them as separate sheets, but the sheets are so arranged that they can be taken out and inserted into the pupil's personal record of laboratory work. (See Chapter 6 for more details of the publications.)

Key to first two lists

Column 1: number of experiment in *Teachers' guide I*

Column 2: ES1, ES2, etc., refer to the pupils' book, *Experiment sheets I*

P indicates an experiment for pupils for which no *Experiment sheet* is supplied

T indicates that the experiment is to be carried out by the teacher

Column 3: title of experiment

Alternative A

A1.1	ES1	Production of pure salt from rock salt
A1.2a	ES2	Boiling ink
A1.2b	T	The distillation of ink
A1.2c	T	How to measure boiling points
A1.2d	ES3	The chromatography of ink
A1.3	ES4	An attempted separation of the green colouring matter in plants by chromatography
A1.4a	T	To prepare solutions of some coloured substances from natural sources
A1.4b	ES5	Using some other coloured substances from plants
A1.5	ES6	Fractional distillation of crude oil
A1.6a	ES7	How can metals be obtained from rocks? (Part 1)
A1.6b	ES7	How can metals be obtained from rocks? (Part 2)
A2.1	ES8	What happens to some chemicals when they are heated?

A2.2	ES9	Investigating the effect of heat on copper sulphate crystals in more detail
A2.3	ES10	To find out if there is any change in weight when certain substances are heated
A2.4	ES11	Tracking down the matter lost when potassium permanganate is heated
A2.5	T and P	Heating copper away from the air
A3.1	T or P	How much air is used up when copper is heated?
A3.2	T	How Priestley first obtained 'active air'
A4.1	ES12	Burning substances in oxygen
A4.2a	ES13	What happens when a candle burns in air?
A4.2b	T	The products of combustion of a candle
A4.3	T	Investigating changes in weight when a candle burns
A5.1a	ES14	An investigation of some elements
A5.1b	T	More properties of elements
A6.1a	T	The reactivity of magnesium and copper
A6.1b	T	The reactivity of aluminium and iron
A6.1c	ES15	Competition for oxygen between iron, zinc, copper, and lead
A6.2a	ES16	How does carbon compare with the metals in its affinity for oxygen? (Part 1)
A6.2b	ES16	How does carbon compare with the metals in its affinity for oxygen? (Part 2)
A7.1	T	Is water formed in a Bunsen flame?
A7.2	T	Will magnesium remove the oxygen from water?
A7.3	ES17	To investigate the action of steam on iron and zinc
A7.4a	T	Preparing hydrogen
A7.4b	T	Is water formed when hydrogen is burned?
A7.5	T	Can hydrogen be used to reduce metal oxides?
A8.1	ES18	Finding out which substances conduct electricity
A8.2	ES19	To find out whether water or solutions conduct electricity
A9.1	ES20	Trying to get iron from iron ore
A9.2	P	An investigation of malachite
A9.3	ES21	An investigation of another mineral
A10.1a	T	The evaporation of sea water
A10.1b	ES22	What dissolved solids are present in sea water?
A10.2a	P	Electrolysing sea water
A10.2b	T	Electrolysing concentrated sea water
A10.3	ES23	Iodine from seaweed

A10.4a	T	Investigating the halogens
A10.4b	ES24	Competition among the halogens

Alternative B

B1.1	ES25	How can crude alum be purified?
B1.2	ES26	How can crude naphthalene be purified?
B1.3	ES27	Investigating different kinds of water
B1.4a	P	How can we obtain pure water from sea water?
B1.4b	T	An efficient way of purifying sea water
B1.5	ES4	An attempted separation of the green colouring matter in plants by chromatography
B2.1a	T	To prepare solutions of some coloured substances from natural sources
B2.1b	ES5	Using some other coloured substances from plants
B2.2	ES28	To investigate the effect of an alkali on an acid
B3.1	T	Fractional distillation
B3.2	ES6	Fractional distillation of crude oil
B3.3	T	To find out if anything can be separated from air by cooling it
B4.1	ES12	Burning substances in oxygen
B5.1	ES29	Finding out what happens when various substances are heated
B5.2a	T	How to measure boiling points
B5.2b	ES9	Investigating the effect of heat on copper sulphate crystals in more detail
B6.1	ES30	Investigation into substances which conduct electricity
B6.2	T	Finding out if molten substances conduct electricity
B6.3	ES31	To investigate what happens when various solutions conduct electricity
B6.4	ES32	Using electricity for zinc plating
B6.5a	ES33	Getting electricity from a chemical reaction
B6.5b	T	Comparing the voltages from different cells
B8.1a	P or T	Properties of hydrogen and oxygen
B8.1b	T	Making water from hydrogen and oxygen
B8.2a	T	The preparation of some sulphides
B8.2b	T	The preparation of some chlorides
B8.2c	ES34	Some properties of oxides, chlorides, and sulphides
B9.1	ES35	An investigation of some of the chemical changes that take place during burning
B9.2	ES36	An investigation of breathing

B9.3	ES37	An investigation of the changes which take place when iron rusts
B9.3a		Part 1: are both air and water necessary for rusting?
B9.3b		Part 2: which gas in the air is responsible for rusting?
B9.3c		Part 3: confirming that oxygen is the gas in the air which is responsible for rusting
B9.4	ES38	Investigating the composition of rust
B10.1a	P or T	Will carbon turn copper oxide to copper?
B10.1b	T	Will hydrogen turn copper oxide to copper?
B10.3	ES39	More competition among the elements
B10.4a	T	The appearance of sulphur and the halogens
B10.4b	ES40	Competition among some non-metallic elements

Experiment sheets

This list gives the number of the *Experiment sheets* in the first column, and the numbers of the experiments in the *Teachers' guide* to which they correspond in the second column, followed by the title of the experiment.

ES	Teachers' guide I		
1	A1.1		Production of pure salt from rock salt
2	A1.2a		Boiling ink
3	A1.2d		The chromatography of ink
4	A1.3	B1.5	An attempted separation of the green colouring matter in plants by chromatography
5	A1.4b	B2.1b	Using some other coloured substances from plants
6	A1.5	B3.2	Fractional distillation of crude oil
7	A1.6a, A1.6b		How can metals be obtained from rocks?
8	A2.1		What happens to some chemicals when they are heated?
9	A2.2	B5.2b	Investigating the effect of heat on copper sulphate crystals in more detail
10	A2.3		To find out if there is any change in weight when certain substances are heated
11	A2.4		Tracking down the matter lost when potassium permanganate is heated
12	A4.1	B4.1	Burning substances in oxygen
13	A4.2a		What happens when a candle burns in air?
14	A5.1a		An investigation of some elements
15	A6.1c		Competition for oxygen between iron, zinc, copper, and lead
16	A6.2a, A6.2b		How does carbon compare with the metals in its affinity for oxygen?

17	A7.3		To investigate the action of steam on iron and zinc
18	A8.1		Finding out which substances conduct electricity
19	A8.2		To find out whether water or solutions conduct electricity
20	A9.1		Trying to get iron from iron ore
21	A9.3		An investigation of another mineral
22	A10.1b		What dissolved solids are present in sea water?
23	A10.3		Iodine from seaweed
24	A10.4b		Competition among the halogens
25		B1.1	How can crude alum be purified?
26		B1.2	How can crude naphthalene be purified?
27		B1.3	Investigating different kinds of water
28		B2.2	To investigate the effect of an alkali on an acid
29		B5.1	Finding out what happens when various substances are heated
30		B6.1	Investigation into substances which conduct electricity
31		B6.3	To investigate what happens when various solutions conduct electricity
32		B6.4	Using electricity for zinc plating
33		B6.5a	Getting electricity from a chemical reaction
34		B8.2c	Some properties of oxides, chlorides, and sulphides
35		B9.1	An investigation of some of the chemical changes that take place during burning
36		B9.2	An investigation of breathing
37		B9.3	An investigation of the changes which take place when iron rusts
38		B9.4	Investigating the composition of rust
39		B10.3	More competition among the elements
40		B10.4b	Competition among some non-metallic elements

Appendix 4

16 mm films and BBC television science programmes

Few references were made to 16 mm films in the first edition because of the development of a large number of 8 mm silent film loops. The great advantage of film loops is that they can be fitted into the fabric of the lesson, with the minimum of fuss, at the correct moment, and leave the instruction entirely in the hands of the teacher. Yet 16 mm films, although inherently less flexible in use than film loops, can be of great value, and can take the pupil out of the laboratory and into the world of industry more effectively than is possible on the silent film loop with its small screen.

It should be remembered that 16 mm films can be used flexibly to some extent; there is no need to show all the film, and of course the sound can be turned down to allow the teacher to give his own commentary. BBC Schools television programmes can be used even more flexibly, as they may be recorded on videotape and may be edited by the school as required. The programmes may be recorded when they are transmitted by the BBC, and then played back. Because of the present copyright laws, recorded programmes should be erased one year after they are recorded.

The use of films and BBC Schools television programmes requires considerable forethought. Films must be ordered as far in advance as possible (sometimes several months) and the BBC programmes must be recorded when they are transmitted. Some BBC programmes are now available from the BBC on hire. A list is available from BBC TV Enterprises, Villiers House, Ealing, London W5. The following table has been prepared as an aid to planning.

Note. The supply of films is constantly changing as organizations update their film lists. Therefore, although the films below have been checked, some may have been withdrawn by the time the teacher applies for them. Teachers are therefore advised to make sure, well in advance, that the film they want is still available, and they may find that they can borrow a more up-to-date film instead.

Films and BBC television programmes arranged in course order

IA	IB	Title	Sponsor	Distributor
A1.5	B3.2	Refining	British Petroleum	Petroleum Films Bureau
A1.5	B3.2	Oil	Shell	Petroleum Films Bureau
A1.5	B3.2	North Sea Strike	Esso	Petroleum Films Bureau
A1.5	B1.2	Coal age?	BBC	*BBC
A1.6	B10.5	Treasure Trove	Imperial Chemical Industries	ICI Film Library
A2.5		Exploring chemistry†		Unilever Film Library
A3.3		History of the Discovery of Oxygen	Imperial Chemical Industries	ICI Film Library
A3.3		Oxygen	BBC	*BBC
A3.3	B3.3	O for Oxygen	British Oxygen	British Oxygen Company
	B3.3	Oxygen in Steelmaking	British Oxygen	British Oxygen Company
	B3.3	Air	Imperial Chemical Industries	ICI Film Library
A4	B4	Combustion		Educational Foundation for Visual Aids
A4	B9.1	Burning	BBC	*BBC
A4	B4	Fire chemistry	Ministry of Aviation	Central Film Library
A4	B4	The air, my enemy	Gas Council	Viscom Film Library
A9.1		The Big Mill	Colvilles (now part of British Steel Corporation)	British Steel Corporation
A9.1	B10.2	Study in Steel	British Steel Corporation	Viscom Film Library
	B10.2	Making Iron	British Steel Corporation	Viscom Film Library
A9.1	B9.3	Corrosion	BBC	*BBC
A9.3		Limestone in Nature	Imperial Chemical Industries	ICI Film Library
A10	B10	Chlorine	Imperial Chemical Industries	ICI Film Library

*These programmes are transmitted by the BBC in the Exploring Science Series. This series is being continually updated, and these titles will not be transmitted indefinitely. The programmes may be recorded on videotape as described on page 246.

†This film is for teachers, not pupils.

Brief descriptions and further details

The following films are arranged by distributor.

Central Film Library

The following film is available on hire from the Central Film Library, Government Building, Bromyard Avenue, London W3.

Fire Chemistry (Ministry of Aviation, 1966) 30 minutes, sound, colour.

Though made to train those in the fire services, this film is an excellent introduction, for pupils of this level, to the nature of fire. While showing the different types of fire and how they may be fought, the film explains clearly that fire needs fuel, heat, and oxygen. There are also some excellent laboratory shots. These illustrate the different types of heat transmission and show a set of experiments explaining flash point and ignition temperatures.

Educational Foundation for Visual Aids

The following film is available on hire from the Educational Foundation for Visual Aids, National Audio Visual Aids Library, Paxton Place, Gipsy Road, London SE27 9SS.

Combustion (1964) $15\frac{1}{2}$ minutes, sound, colour.

This is very useful support material, demonstrating the ideas of fuel, oxidation, kindling temperature, and spontaneous combustion. The film uses animation to show the molecular action of a burning fuel.

British Steel Corporation

The following film is available on free loan from General Steels Division, Film Library, British Steel Corporation, Colville House, 120 Bothwell Street, Glasgow G2.

The Big Mill (Colvilles, now part of British Steel Corporation, 1961) 27 minutes, sound, colour.

This is the story of the giant steel strip mill at Ravenscraig, Lanarkshire, from the building of it to the production of steel in it. The film shows the importance of careful temperature checks and analysis of samples of steel, and there are exciting shots of a huge bath of bubbling steel, and of a thick slab of steel rolled at high speed to become a thin coil four miles long.

British Oxygen Company

The following British Oxygen films are available on free loan from British Oxygen Company, Film Library, 42 Upper Richmond Rd West, London SW14 8DD.

O for Oxygen (British Oxygen Company, 1960) 23 minutes, sound, colour.

This award-winning film explains simply and clearly how the air is separated into its components. After briefly covering the uses of nitrogen and the noble gases, the film concentrates on the world-wide use of oxygen in steelmaking, cutting ships' plates, welding, and saving lives. Teachers may wish to stop the film before the section on welding, as this is rather detailed for this stage.

Oxygen in Steelmaking (British Oxygen Company, 1960) 20 minutes, sound, colour.

Animation is important in this film in explaining the various methods of steelmaking in Great Britain, Sweden, and Austria, and also in explaining how oxygen is made. The oxygen-producing plant at Margam in South Wales is shown. Though shorter than 'Study in Steel' (see page 246) this film has a similar content.

ICI Film Library

The following ICI films are available on free loan from ICI Film Library, Thames House North, Millbank, London SW1P 4QG.

Air (1969) 14 minutes, sound, colour.
After establishing that air has weight, the film explains how an aneroid barometer works and how air pressure affects pilots and divers. It then deals with the chemistry of extracting gases from the air, and how they are used.

Chlorine (1963) 15 minutes, sound, colour.
This illustrates the importance of chlorine for water purification, industrial solvents, aerosol propellants, and so on. With animated diagrams and live action it explains the large-scale manufacture of chlorine by the mercury-cell process.

History of the Discovery of Oxygen (1947) 16 minutes, sound, colour.
This is best shown after pupils have completed Topic A3. Starting with the alchemists, it traces the development of ideas on burning, via the phlogiston theory, to the work of Scheele, Priestley, and Lavoisier.

Limestone in Nature (1948) 10 minutes, sound, black and white.
This is a short film explaining how marine creatures were formed, millions of years ago, into chalk, limestone, or marble. The second part shows how limestone is dissolved by rain water and carbon dioxide from the atmosphere.

Treasure Trove (1956) 19 minutes, sound, black and white.
The treasure trove in this case is rocks, and in particular salt, anhydrite, and limestone. The film is an old one and does not include petrochemicals. Nevertheless, it is still worth showing.

Petroleum Films Bureau

The following British Petroleum, Shell, and Esso films are available for a very small charge to cover outward postage from the Petroleum Films Bureau, 4 Brook Street, London W1Y 2AY.

North Sea Strike (Esso, 1967) 22 minutes, sound, colour.
'North Sea Strike' is about looking for gas, from the seismic surveys to the drilling and the dramatic strike as gas rushes up from two miles underneath the sea bed. While capturing the noisy atmosphere of the drilling rig, the film also shows the value of the chemists who use chromatography to analyse samples from the sea bed.

Oil (Shell, 1971) 18 minutes, sound, colour.
'Oil' replaces 'Introduction to Oil' (1962). It is an excellent introduction to the subject – how oil was formed, how it is found, and how it is transported and used. The film demonstrates our dependence on oil for so many things besides fuel, such as paints, rubbers, fertilizers, and plastics. Particularly effective is the juxtaposition of shots of fractionating columns with shots of the same process, in miniature, in a laboratory.

Refining (British Petroleum, 1968) 18 minutes, sound, colour.
This film has an original and humorous approach to the problem of explaining such a complex subject as oil refining techniques. A cartoon figure detaches himself from a party touring a refinery and, commenting ironically on the jargon used by the party guide, explains clearly and simply what the guide really means.
'Refining' gained a festival award for its imaginative approach.

Unilever Film Library

The following film is available on free loan from Unilever Film Library, Unilever House, London EC4P 4BQ.

Exploring Chemistry (1967) 35 minutes, sound, colour.
This was produced for the Nuffield Science Teaching Project to show an example of the investigational approach to chemistry. The film starts by examining the learning process through early childhood up to school age, and then the scene shifts to a chemistry

laboratory in a comprehensive school where pupils are encouraged to adopt a questioning attitude to their subject.

Viscom Film Library

The following British Steel Corporation and Gas Council films are available on free loan from Viscom Film Library, 16 Paxton Place, London SE27 9SS.

Making Iron (British Steel Corporation, 1970) 30 minutes, sound, colour.
This is available as three films: 'The Handling and Preparation of Raw Materials' 13½ minutes (showing open cast mining and the use of limestone, coke, and air in obtaining iron from its ore); 'The Blast Furnace' 8½ minutes (explaining by diagrams and animation the construction of a furnace and the chemical reactions that take place, and showing a furnace being tapped); and 'What comes out of the Blast furnace' 8 minutes (showing how the by-products, gas and slag, are cleaned).

Alternatively, the three parts are available as one thirty-minute film.

Study in Steel (British Steel Corporation, 1968) 27 minutes, sound, colour.
The first half of this film, up to the description of the uses of steel, is the most relevant at this stage. The film shows the different methods of steelmaking, including the relatively new process of making spray steel: jets of molten iron, oxygen, and lime converge to form 'instant steel'.

The air my enemy (The Gas Council, 1970) 25 minutes, sound, colour.
Teachers may feel that pupils have seen enough on television about pollution. However this award-winning film takes a carefully considered look at this crucial problem, and in particular at air pollution. The film explains that smoke in London has been very much reduced, but goes on to demonstrate very effectively the damage being done by sulphur dioxide.

BBC television programmes on chemistry

The BBC transmits a series of twenty-minute television programmes in the Exploring Science series, including 'Corrosion', 'Oxygen', 'Burning', and 'Coal age?'.

The series, which is being continually updated, aims to relate the work going on in the laboratory to real problems and situations in the outside world. An excellent set of teacher's notes is available from the BBC. The series can of course be viewed by tuning in at the advertised times, but the most effective and flexible use of the programmes can be made by recording the programmes on videotape. Teachers who do not yet have access to videotape recording facilities may care to:
1. Contact their local Science Advisor, Audio-Visual Aids Officer, or Science Inspector.

2. Write to the National Committee for Audio-Visual Aids in Education, 33 Queen Anne Street, London W1.
3. Many U.D.E.s and Colleges of Education have video facilities.

Corrosion
This programme deals with some of the chemistry of corrosion, and takes as its subject the motor car. It shows experiments illustrating such things as the tendency of mild steel to corrode, the corrosion where two metals touch, and the increase in corrosion when sulphur dioxide and salt are present. The programme then deals with how manufacturers try to protect cars from corrosion, including using zinc-coated steel in expensive cars. The advantage of this is illustrated in a final demonstration that when zinc and mild steel strips are placed in a beaker of water and connected, the zinc rather than the steel corrodes.

Oxygen
'Oxygen' deals with the experiments by Priestley, Scheele, and Lavoisier that led to the discovery and naming of oxygen nearly 200 years ago. The programme also points out, however, that it was not until William Hampson developed a suitable machine that large quantities of oxygen could be made cheaply, which has led to millions of tonnes of oxygen being used throughout the world every year.

Burning
In explaining the idea of burning, this programme gives some useful information on fires in the home. It explains with experiments that different substances burn at different rates, and that burning is also affected by the mixture of the reagents. The best mixture is a molecular mix of liquid or gases, demonstrated by the difference between the rate of reaction of one balloon filled with hydrogen and another filled with a mixture of hydrogen and oxygen.

Coal age?
After an animated film of one theory of how coal was formed, 'Coal age?' shows a modern mine in operation. It tells how the uses of coal have changed over the years, partly because of the introduction of smokeless zones and partly because of discoveries of new uses for coal products – such as using pitch (from coke manufacture) as a damp course on modern roads and motorways.

Appendix 5

School and public examinations

The impact of examinations on learning and teaching

For many years now, public examinations have had a considerable influence on what goes on in schools. This influence can be a good one; it can also result in serious harm as numerous complaints suggest. The Nuffield proposals are based on two observations and on the conclusions that can be drawn from them.

1. The influence of examinations, good or bad, is strongly felt in the years preceding the public examinations. For many pupils, the public schools' common entrance examination, the O-level G.C.E., the C.S.E., or the O.N.C. has become their goal, and it is unquestionably our duty to see that they are helped and guided in their endeavours to reach this goal. This responsibility naturally results in all examinations – class tests, end-of-term examinations, and end-of-year examinations – reflecting to a lesser or greater degree the type of demand made in the public system. In turn, therefore, it affects what we teach, the approach we take, and how the pupils learn.

As a result of this pattern, when we discuss examinations in our proposals we are not thinking only of public examinations. We are thinking of all those occasions on which pupils are confronted with trial and assessment of progress.

2. What pupils find they must do, term in, term out, to achieve success, they naturally regard as indicative of what we teachers consider to be important and valuable. In this they are not unreasonable, although they may be wrong. However, they also unthinkingly learn to regard it as indicative of what 'chemistry' or 'history' or 'mathematics' *is*. In this they can be seriously misled and those who go on with their studies in science find it almost impossible to unlearn this interpretation. Recent comments from the universities show this is true of many abler children.

It follows that both internal and external examinations ought to encourage, as well as assess, those abilities we want to develop in children.

Abilities which examinations on Stage I work should encourage and assess

It will be recalled from Chapter 3, page 14, that the most important general objectives for Stage I – that is, the principal abilities to be developed – are:

1. Facility in recalling information and experience.
2. Skill in handling materials, manipulating apparatus, carrying out instructions for experiments, and making accurate observations.
3. Skill in devising an appropriate scheme and apparatus for solving a practical problem.
4. Skill in handling and classifying given information.

5. Ability to interpret information with evidence of judgment and assessment.

Without giving opportunity for misunderstanding it is very difficult to say to what extent ability (1) should be demanded. *All* questions are based on the memory of experiences and of occasions of learning skills. *All* questions, therefore, demand facility to recall. However, even in Stage I, and certainly later, we feel that ability to recall, if it is unaccompanied by any evidence of understanding, should certainly not result in a candidate being graded higher than one of the lowest two points on a five-point scale. It follows that an examination paper must not consist chiefly of questions which merely demand regurgitation of information or notes dictated on possible topics.

Practical examinations are sometimes used in an attempt to test ability (2), but these are not really appropriate during Stage I and are also subject to adverse criticism at later Stages. Written questions can be devised to test ability (3) and these go some way towards testing practical skills (2), but written questions alone cannot satisfactorily test the skill which comes from the actual handling of apparatus and chemicals.

To test ability (4), the pupil must be given a good deal of qualitative and quantitative information. Either this information is contained in the question or the pupil is allowed to use a book of data in the examination.

Other objectives mentioned in Chapter 3, of less importance at the Stage I level, are:

6. Ability to apply previous understanding to new situations and to show creative thought.
7. Competence in reporting on, commenting on, and discussing matters of simple chemical interest.
8. Awareness of the place of chemistry among other school subjects and in the world at large.

These, too, must be considered when examinations are being devised.

Questions which test such qualities as judgment, understanding, and creative thought, (5) and (6), are not easy to set, nor are the answers easy to mark, for they are not readily confined to a rigid marking schedule. The pupil who shows commendable original thought may well give an answer which is very different from the model answers which the examiner had in mind.

Types of question

Three main types of question have been found useful at Stage I level. They are:

1. Multiple choice questions
2. Structured questions
3. Free response questions

In a multiple choice question the pupil is asked to pick the correct answer from a number – usually four or five – of possible answers.

Much work in science and, for that matter, day-to-day living, consists of making a choice between a small number of possible alternatives. It seems reasonable, therefore, that this activity should be reflected in some examination questions. A good multiple choice question requires real understanding and judgment for it to be answered correctly, and it also requires much skill in setting if it is to be unambiguous and free from unsuspected alternatives. One advantage which this type of question has over all other types is that it is very easy to mark.

Structured questions consist of a number of short parts which lead on from one to another. Space in which to write the answers is usually provided after each part. The parts are simple to begin with, but may get more difficult towards the end of the question.

The free response or essay-type question includes any question which requires a piece of continuous prose as its answer. It can be difficult to answer, because the pupil must recollect all the relevant information and arrange it logically and grammatically. On the other hand, they can be dangerously easy to set; for example, it is easy to ask, 'Describe the Solvay process', but it is difficult for the pupil to decide what information the examiner expects in return. This is the sort of question which leads to dictated notes and their subsequent regurgitation in examinations.

Each of these types of question can be used to assess any of the abilities mentioned on pages 248–249, although certain abilities are best tested by certain types of question. The multiple choice method of questioning is very flexible, but is perhaps most suited to testing abilities (1) and (4). Structured questions can be devised to test all abilities, although as previously mentioned, no written question is really suitable to test ability (2). Essay questions are almost essential when testing ability (7).

Setting examination questions

The art of teaching is, in large measure, the art of asking questions; and, if a new approach to the teaching of chemistry is to be successful, teachers will find it necessary to devise questions which are in tune with the new approach. This is not easy; each question requires much thought, discussion, and trial before it can be said to test those abilities which it claims to test. But it is worth the effort and teachers will find that devising such questions can be useful and stimulating.

These questions are nearly always longer than the conventional ones, although the answers may be shorter. They require more time to read and comprehend, so if there must be a time limit for examinations of this type, it should be a generous one. And, for at least the younger pupils, it will be found that many of the questions will have to be broken down into several parts; if a pupil in his early days of chemistry is asked to make a judgment or express an opinion, the question must be so devised that he is doing so on one or two specific matters unclouded by side issues and irrelevant information.

Most of the sample questions, particularly the short answer and multiple choice type, can be tied to marking schemes as clearly defined as those in present use. But, if we are to discover in some of our pupils a spark of something a little out of the ordinary, there must be some questions not rigidly bound to a model answer; Q21 is an example. The answers to such questions will have to be marked by impression and it will require experienced chemistry teachers to do so. It should be possible for teachers to mark consistently in this manner provided that they do not have to use more than a five-point basis (ABCDE or 43210).

Above all, questions should be set because they reflect the spirit of the course and not because they are easy to set and easy to mark.

Sample questions to illustrate the Nuffield proposals

There now follows a short selection of questions suitable for use at Stage I level. They are grouped by question type (multiple choice, structured, and essay) and an indication of the abilities tested is given at the end of each question.

Multiple choice questions

1. Carbon at red heat will remove oxygen from both copper oxide and zinc oxide but not from magnesium oxide. Zinc will remove oxygen from copper oxide. On this evidence, which of the following is the order of activity of the three metals, putting the most reactive metal first?

(a) Zinc – magnesium – copper.
(b) Copper – zinc – magnesium.
(c) Magnesium – copper – zinc.
(d) Magnesium – zinc – copper.

Ability tested: 4
Answer: (d)

2. Suppose you wanted to make an electrical connection between the wire A and the electrode B which is to be used for the electrolysis of an aqueous solution. Which one of the following would you put into the glass tube C?

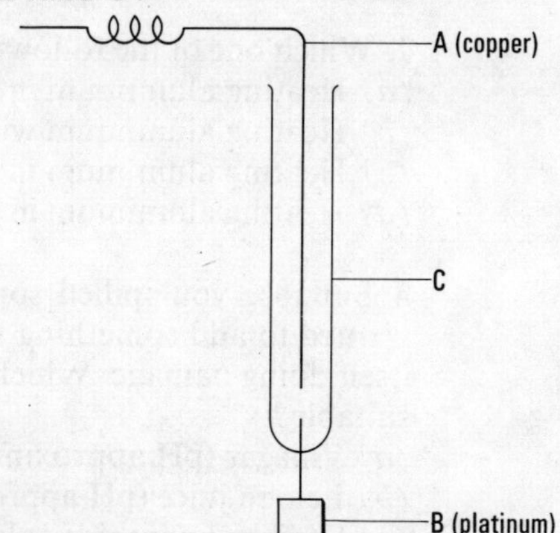

(a) Mercury.
(b) Molten sulphur.
(c) Pure water.
(d) Powdered sulphur.

Ability tested: 3
Answer: (a)

3. Barium is a metallic element which combines with the non-metallic element chlorine to form a compound barium chloride. When barium chloride is molten it conducts electricity. Which one of the following would be formed at the cathode?
(a) Barium.
(b) Chlorine.
(c) Nothing.
(d) Barium and chlorine.

Ability tested: 4
Answer: (a)

4. A certain gas extinguished a candle flame immediately; it had no action on limewater and copper did not change when heated in the gas. Which one of the following could it be?
(a) Air.
(b) Nitrogen.
(c) Carbon dioxide.
(d) Oxygen.

Ability tested: 5
Answer: (b)

5. Four substances, A, B, C, and D, were heated in oxygen; they behaved as given below. Which one of them could not have been an element?
(a) Melted, but did not form a new substance.
(b) Burned to form sulphur dioxide and water.
(c) Burned to form a single oxide.
(d) Did not melt and did not form a new substance.

Ability tested: 5
Answer: (b)

6. Calcium carbonate is a solid which is insoluble in water. Suppose you were given a suspension consisting of 1 g of powdered calcium carbonate in suspension in 200 cm^3 of water. Which one of the following methods would you use to obtain water which no longer had any calcium carbonate in it?
(a) Chromatography.
(b) Filtration.
(c) Evaporation.
(d) Distillation.

Ability tested: 3
Answer: (b)

7. Which one of the following could not produce aluminium oxide?
(a) Heating aluminium in air.
(b) Heating aluminium with iron oxide.
(c) Heating aluminium in nitrogen.
(d) Heating aluminium in steam.

Ability tested: 1
Answer: (c)

8. Suppose you spilled some sulphuric acid on the floor and you wanted to add something which would neutralize the acid without itself doing damage. Which one of the following would be the most suitable?
(a) Vinegar (pH approximately 4).
(b) Lemon juice (pH approximately 5).
(c) Sodium hydroxide solution (pH approximately 14).
(d) Sodium hydrogen carbonate solution (pH approximately 8).

Ability tested: 3
Answer: (d)

9. The electrolysis of a certain liquid resulted in the formation of hydrogen at the cathode and chlorine at the anode. Which one of the

following could it be?
(a) A solution of sodium chloride in water.
(b) A solution of sulphuric acid in water.
(c) Pure water.
(d) A solution of copper chloride in water.

Ability tested: 5
Answer: (a)

10. When crystals of Epsom salts (hydrated magnesium sulphate) are stirred in water they disappear and a clear liquid remains. The crystals can be re-formed by evaporating some of the water. Which one of the following best describes what happens when the crystals are stirred with water?
(a) The crystals melt.
(b) The crystals dissolve.
(c) The water and the crystals react to form a new substance.
(d) The crystals evaporate.

Ability tested: 1
Answer: (b)

11. A certain solid X was heated in a stream of a gas called ammonia (a compound of nitrogen and hydrogen). X changed into another solid which was a good conductor of electricity and at the same time water and nitrogen gas were formed. Which one of the following is the most reasonable conclusion?
(a) X was the oxide of a metallic element.
(b) X was the oxide of a non-metallic element.
(c) X was a metallic element.
(d) X was a non-metallic element.

Ability tested: 5
Answer: (a)

12. Below are ten questions. Find the answer to each by studying the information in the table.

Substance	Electrical conductance	Melting point/°C	Boiling point/°C	Effect of heating in air
A	Good when solid or liquid	97	889	Burns to form a single oxide which forms an alkaline solution in water
B	Non-conductor	113	444	Burns to form a single oxide which forms an acidic solution in water
C	Non-conductor	5	80	Burns to form carbon dioxide and water
D	Non-conductor when solid, good conductor when molten	800	1413	Melts; no new substance formed

(a) Which substance would be a liquid at room temperature?
(b) Which substance would boil if put into a tube surrounded by boiling water?
(c) Which substance stays liquid over the greatest range of temperature?
(d) Which substance could be a metallic element?
(e) Which substance could be a non-metallic element?
(f) Which substance contains carbon?
(g) Which substance could be sodium chloride?

(Continued)

Ability tested (all parts): 4
Answers: (a) C, *(b)* C, *(c)* A, *(d)* A, *(e)* B, *(f)* C, *(g)* D, *(h)* D, *(i)* C, *(j)* B

(h) Which substance when heated in air would not change in weight?
(i) Which substance most nearly resembles petrol?
(j) Which substance could be sulphur?

Structured questions

13. The apparatus drawn below was used to make hydrogen and to investigate the action of hydrogen on the oxide of a metal.

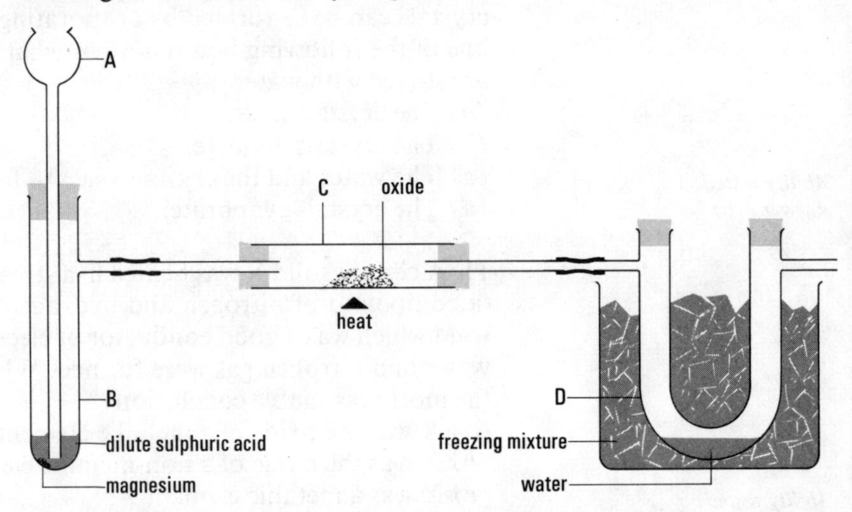

The tube containing the oxide (C) was weighed; hydrogen was then made in the tube B and passed over the oxide which was heated when all the air had been swept from the apparatus. Water collected in the tube D. The tube C was cooled and reweighed; it was found to weigh less than it did before the experiment started.
(a) Name the piece of apparatus labelled A.
(b) Explain the loss in weight of the contents of the tube C.
(c) A student pointed out that the formation of water in the tube D did not prove that water had been formed by the reaction in the tube C. What changes would you make in the apparatus to make certain that the water obtained did result from the reaction in C?
(d) How would you show that the liquid which collected in D consisted, at least in part, of water?
(e) State one further piece of evidence which would indicate that the liquid in D was pure water.
(f) When the reaction started in C the oxide glowed red hot. The Bunsen burner was removed and the oxide continued to glow red until the reaction was complete. What does this indicate about the reaction which took place in C?

Abilities tested *(a)* 1, *(b)* 5, *(c)* 6, *(d)* 3, *(e)* 3, *(f)* 5

Marking scheme
(a) 1 for thistle funnel.
(b) 2 – 1 – 0. Full marks for the candidate who realizes that the oxide has lost oxygen and that the hydrogen has taken it. 1 mark for loss of oxygen alone.
(c) 2 – 1 – 0. For anything which could possibly dry the hydrogen – 2 marks. If the candidate realizes that the hydrogen must be dry but cannot suggest a method of drying it – 1 mark.
(d) 2 – 1 – 0. Full marks for the name of the substance which will show the presence

of water and also the change which would be observed in the substance. 1 mark if the observation is omitted.
(e) 1. Any physical constant of water will do; boiling and freezing point are the most likely.
(f) 2 – 1 – 0. Full marks for exothermic reaction or heat given out. For vigorous reaction alone – 1 mark.

14. Read the following passage and then answer the questions.

Aluminium and electricity

The method of using an electric current to take compounds apart is called electrolysis. Electrolysis is widely used in the chemical industry. It is used, for example, in the production of hydrogen, chlorine, and sodium, and in the refining of gold, silver, and copper. Before the industrial use of electrolysis at the end of the last century, aluminium metal was even rarer than gold, despite the fact that about eight per cent of the Earth's crust consists of aluminium and it is potentially the most common metal. But the aluminium is difficult to obtain because, in nature, nearly all of it is strongly combined with oxygen gas to form alumina. There are several alumina ores and the commonest is called bauxite. First attempts to split the alumina with an electric current were a failure because electrolysis can take place only in a liquid and the solid alumina has a very high melting point. However, it was discovered in 1886 – by Charles Hall in America and by Paul Heroult in France – that alumina would dissolve in a mineral of low melting point called cryolite. Cryolite is a glassy-white mineral (named after the Greek for frost, *kruos*), which is found in large quantities in Greenland. The discovery that alumina would dissolve in molten cryolite has led to the production of aluminium on a large scale using electrolysis. It needs a lot of electricity to do this, and therefore the aluminium can be obtained economically only where cheap electric current is available. Also, the aluminium ore must be very pure: ordinary garden soil contains 6–8 per cent aluminium in the form of aluminium compounds, but it is much too expensive to extract.

(a) Name two elements of which bauxite is largely composed.
(b) Why is aluminium frequently extracted from sources near hydro-electric power stations?
(c) Explain why aluminium was rarer than gold a hundred years ago even though aluminium compounds form such a large proportion of the Earth's crust.
(d) How did the discovery of cryolite lead to the large-scale production of aluminium?
(e) Name one non-metallic element which is made by electrolysis in industry.
(f) What does the passage you have just read indicate about the melting point of alumina?

Abilities tested *(a)* 4, *(b)* 5, *(c)* 5, *(d)* 5, *(e)* 1, *(f)* 5

Marking scheme *(a)* 1 each for aluminium and oxygen. A candidate may know bauxite is hydrated, so hydrogen should be allowed.
(b) 2 – 1 – 0. 1 mark each for any relevant suggestion: for example, electricity is neces-

sary, it is convenient and/or cheaper to have the generating station nearby, the cost of hydro-electric power is lower, it brings employment to areas where industrial employment is usually scarce.
(c) 2 – 1 – 0. Give 1 mark if candidate has realized that aluminium is more difficult to extract or is not found native; 1 mark if a reason is given – for example, that electricity is necessary, aluminium compounds are more stable, it is difficult to melt aluminium oxide, or aluminium is firmly bound to oxygen.
(d) 2 – 1 – 0. If a candidate shows that he is aware that the liquid state is necessary for electrolysis and that cryolite made this possible he should get full marks. Any two relevant points – for example, aluminium oxide is difficult to melt but it dissolves in molten cryolite – should get full marks.
(e) 1 mark for any non-metallic element so produced – allow hydrogen.
(f) 1 mark for high melting point or difficult to melt.

15. Here are the results of an investigation into a liquid which can be called X.
Read them and then answer the questions which follow.
X was found to be a good conductor of electricity; it had a pH of 7, a boiling point of 102 °C, and a freezing point of −7 °C.

Some of X was put into a weighed beaker which was then reweighed. The beaker was put into an oven at a temperature just below 100 °C until all the liquid evaporated leaving a solid which was called Y. The beaker was reweighed.

Weight of beaker	26.05 g
Weight of beaker plus X	38.55 g
Weight of beaker plus Y	27.65 g

Some of X was distilled and a liquid called Z was collected. Z was found to be a bad conductor of electricity; it had a pH of 7, a boiling point of 100 °C, and a freezing point of 0 °C.

(i) Which of the following conclusions are justified in the light of these results (put a ring round each of the conclusions which you choose)?
(a) X is a pure substance.
(b) X is a solution.
(c) Y is an electrolyte.
(d) Y is not an electrolyte.
(e) X has a lower boiling point than Z.
(f) X has a lower freezing point than Z.

(ii) Name one liquid which Z could be.

(iii) Using the results given above, calculate
(a) the weight of X which was put into the beaker;
(b) the weight of Y which remained in the beaker;
(c) the weight of liquid which had been lost by evaporation.

(iv) What weight of X would you evaporate to obtain 3.20 g of Y?

(v) What is the maximum weight of Z which could be obtained by distilling 6.25 g of X?

Abilities tested *(i)* 5, *(ii)* 5, *(iii)* 4, *(iv)* 4, *(v)* 4

Marking scheme *(i)* 1 mark each for *(b)*, *(c)*, and *(f)*. Total 3. −1 for each wrong choice.
(ii) 1 for water.
(iii) *(a)* ½ for 12.50 g (12.5 g will do); *(b)* ½ for 1.60 g (1.6 g will do); *(c)* 1 for 10.90 g

(10.9 g will do). No mark if units are not stated. (The maximum loss of marks for not stating units should be 1.)
(iv) 1 for 25.00 g (25 g will do). No mark if units are not stated. (The maximum loss of marks for not using units should be 1.)
1 for correct method. A good candidate might do it in his head; give credit. (If both working and answer are consistent with an incorrect answer to *(iii)*, give 2 marks.)
(v) 1 for 5.45 g. Units as above.
1 for correct method. See *(iv)* above.
(If both method and answer are consistent with an incorrect answer to *(iii)*, give 2 marks.)

16. *(a)* From the following information, list, in the order of their reactivity, the four metallic elements mentioned, putting the most reactive one first.
Lead can be obtained by the action of either magnesium or zinc on a lead compound.
Silver can be obtained by the action of lead on a silver compound.
Zinc can be obtained by the action of magnesium on a zinc compound.

(b) Copper is less reactive than lead. Would you expect powdered zinc to react with copper oxide?
If so, what substances would you expect to be formed?

(c) Name three elements which are in the same family of elements (the halogens) as chlorine.

(d) Choose any one of the elements which you have named in answer to part *(iii)* (not chlorine) and state one natural source from which it is obtained and one use to which the element is put.

Source:

Use:

Abilities tested — *(a)* 4, *(b)* 4, *(c)* 1, *(d)* 1

Marking scheme
(a) 2 for magnesium – zinc – lead – silver. 0 marks for anything different.
(b) 1 for yes. 1 each for copper and zinc oxide.
(c) 1 each for fluorine, bromine, and iodine.
(d) 1 for any source and 1 for any use.

17. The following apparatus was used to investigate the action of air on heated copper. The tube containing the copper was weighed before the apparatus was assembled.

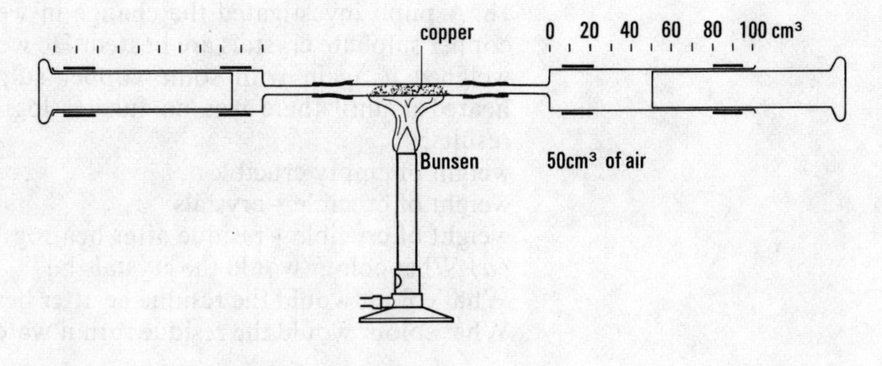

(Continued)

By means of the syringes air was passed backwards and forwards over the heated copper until no further change could be observed. At this point some unchanged copper remained. After allowing the apparatus to cool to its original temperature, the volume of the remaining air was measured and the tube containing the copper was reweighed.

(i) Which one of the following changes would you expect in the weight of the tube?
(a) It would decrease.
(b) It would remain the same.
(c) It would increase.

(ii) Which one of the following volumes would you expect for the final volume of air?
(a) 40 cm^3.
(b) 50 cm^3.
(c) 30 cm^3.
(d) 60 cm^3.
(e) 70 cm^3.

(iii) Why was the apparatus cooled to its original temperature before the final volume of air was measured?

(iv) What effect, if any, would there be on the results of this experiment if more copper had been used?
Why?

(v) What difference in the results would you expect if the syringe had contained air from your lungs instead of normal air?
Why?

(vi) It was found that, if the tube containing the copper was weighed while it was still hot, its weight altered during the process of the weighing. What do you think is a likely explanation of this?

Abilities tested *(i)* 1, *(ii)* 4, *(iii)* 1, *(iv)* 5, *(v)* 5, *(vi)* 5

Marking scheme *(i)* 1. *(ii)* 1. *(iii)* 1 (for any mention of the expansion of a gas on heating or its contraction on cooling). *(iv)* 1 for no effect, 1 for realizing that there was nothing left in the air with which the copper would react. *(v)* 1 for a smaller decrease in volume. 1 for the fact that there is less oxygen in expired air and 1 for explaining that this was because some of the oxygen had been used in the body by a process resembling burning. *(vi)* 1 for the air getting into the tube and 1 for the copper reacting with it. (Maximum 10.)

18. A pupil investigated the change in weight which occurs when copper sulphate crystals are heated. He weighed an empty crucible, weighed it again with some copper sulphate crystals in it, and heated it until there was no further loss in weight. Here are his results:

weight of empty crucible	16.02 g
weight of crucible + crystals	17.52 g
weight of crucible + residue after heating	16.98 g

(a) What colour would the crystals be?
What colour would the residue be after heating?
What colour would the residue turn if water were added to it?

(b) What was the weight of the crystals?
What was the loss in weight of the crystals?

(c) Which one of the following is the most likely explanation of the loss in weight of the copper sulphate?
(i) The volume of copper sulphate has shrunk.
(ii) The copper sulphate has burned.
(iii) The copper sulphate lost something when it was heated.
(iv) The copper sulphate has evaporated.

(d) Suppose you were asked to make 4.8 g of anhydrous copper sulphate. *Using the results given above*, calculate the weight of copper sulphate crystals you would start with.

Abilities tested *(a)* 1, *(b)* 4, *(c)* 1, *(d)* 4

Marking scheme *(a)* 3 × 1. *(b)* 2 × 1. *(c)* 2.
(d) 3 for correct answer (7.5 g), working need not be shown. 2 for incorrect answer if due to an arithmetical error; correct working must be shown. (Maximum 10.)

19. Here are four lists of common substances:
(a) sodium chloride, lead oxide, lead bromide.
(b) alcohol, sugar, naphthalene.
(c) carbon, sulphur, oxygen, nitrogen.
(d) sodium, magnesium, iron, lead.

(i) Which one of these lists consists of metallic elements?

(ii) Which one of these lists consists of non-metallic elements?

(iii) Which one of these lists consists of *compounds* which conduct electricity when they are molten?

(iv) Which of these lists consists of *compounds* which do not conduct electricity when they are molten?

(v) Choose one of the *compounds* which when molten conducts electricity and name the products which are formed in the process.
Name the products chosen:
(a) and *(b)*

Which one of these is formed at the cathode?

(vi) Suppose that you were asked to show that a chemical reaction can produce electricity; draw a diagram of the apparatus and circuit which you would use and label the important parts.

Abilities tested *(i)* 4, *(ii)* 4, *(iii)* 4, *(iv)* 4, *(v)* 4, *(vi)* 1

Marking scheme *(i)* to *(iv)* 1 each. *(v)* 3 × 1.
(vi) Suitable electrodes 1, suitable electrolyte 1, means of detecting current 1. Do not penalize for drawing unless it is so bad as to be misleading. (Maximum 10.)

20. List (1) is of three metallic elements. List (2) is of three non-metallic elements. In both lists the most reactive element is placed first and the least reactive element last.
(1) aluminium, iron, lead
(2) chlorine, bromine, iodine

(Continued)

(i) Which one of the following is most likely to happen?
(a) Aluminium will combine with lead.
(b) Aluminium will combine more vigorously with chlorine than iron combines with chlorine.
(c) Lead will combine with bromine more vigorously than aluminium combines with bromine.
(d) Iodine will combine with iron more vigorously than chlorine combines with iron.

(ii) If you were asked to change iron oxide into iron by heating it with one of the other elements in list (1), which element would you choose?
Give reasons for your answer to *(ii)*.

Abilities tested (i) 4, (ii) 6

Marking scheme (i) 3. (ii) 2 for aluminium.
Aluminium is more reactive than iron 1. Therefore aluminium will combine more readily with oxygen than iron will 1. Therefore aluminium will remove (reduce) oxygen from iron oxide 1. This part could well be marked by impression 3, 2, 1, 0. (Maximum 10.)

Essay questions

21. A gardener was asked to investigate the effect of lime on the growth of sweet peas in a certain sample of soil. He divided the soil equally between two large pots and with the soil in one pot he mixed a weighed amount of lime. He then planted an equal number of seeds in each pot and thereafter gave both groups of plants identical treatment.

Say why you think that he went to the trouble of planting seeds in soil which contained no lime.

Abilities tested 1, 5, 7, 8

22. Suppose that you were camping near a supply of river water and you wished to make a small quantity of distilled water to 'top up' an accumulator. If you had a kettle, one or two bottles, and a primus stove, describe how you would do this and explain what is happening during the process.

Abilities tested 1, 3, 6, 7, 8

23. 'The Haber process for the manufacture of ammonia is helping in the fight against poverty and disease.' In the space allowed below, give your opinion of this statement.

Abilities tested 1, 7, 8

Appendix 6

A brief history of the project

The Nuffield Chemistry Project is one of a number of similar ventures aiming to improve chemistry teaching in various parts of the world. Like many of the others, it arose from a desire for a new attitude and a modern approach to chemistry teaching.

It started in 1957 when the Science Master's Association (now the Association for Science Education) published a policy statement setting out proposals about what part science subjects should play in the educational system as a whole. In 1961 separate panels (with SMA and AWST members) reported on syllabuses and suggested teaching methods for chemistry, physics, and biology.

By the end of 1961 it was clear that, to implement these proposals, outside assistance was needed. This came from The Nuffield Foundation which undertook to sponsor and administer a series of projects to develop new science curricula for schools. In all, £430 000 was provided. The proposals of the SMA–AWST chemistry panel were taken as the starting point for the Nuffield Chemistry Project.

Groups of teachers were seconded, and the group responsible for the chemistry programme assembled in September 1962. Initially the Headquarters Team, as it came to be called, numbered four, but additional members were recruited as necessary. The organizer for chemistry, Mr H. F. Halliwell (now Professor Halliwell), directed the team and a Consultative Committee under the Chairmanship of the late Professor R. S. Nyholm, F.R.S., gave support and advice.

A few weeks later, the Consultative Committee approved a series of guiding memoranda which dealt with such matters as the approach to and the content of a teaching programme, the publications and other resources such a programme would need, the arrangements to ensure that public examinations would be in sympathy with the intention and spirit of the teaching programme, the recommended nomenclature and units, and a provisional timetable for the different stages of the project.

It was decided to illustrate the proposals by a detailed teaching course known as the 'Sample Scheme'. In fact for Stage I alternative 'Sample Schemes' were produced.

In December 1962 twelve teachers were asked to give help and advice on a teachers' guide. They produced a large part of the *Handbook for teachers*, and made an invaluable contribution to the project as a whole.

With the cooperation of the Shell Chemicals Co. Ltd, Mr H. P. H. Oliver was seconded to act as editor of the series of Background

Adaptations of Nuffield Chemistry are being used in many developing countries of the world, especially in East Africa (Kenya, Tanzania, and Uganda) and in Malaysia. This photograph shows pupils in Tanzania preparing and collecting hydrogen (Experiment A7.4a).
Photograph, J. A. Kent.

Books, and detailed work started on these in March 1963. Also in early 1963, with very generous financial assistance from Shell International Ltd, the Shell Film Unit under Mr D. Segaller started to plan and produce the film loops.

At the same time thirty teachers were appointed to develop experiments for the Sample Scheme. They did a considerable amount of original work. Much of this is in the Sample Scheme, and all of it is in *Collected experiments*.

The trials

Initial trials of portions of the Sample Scheme were carried out in twelve schools in the Autumn Term of 1963, setting the pattern for more widespread trials later. During this time the late Dr F. H. Pollard of the University of Bristol led a group to investigate ways to examine, test, and assess pupils' progress.

In January 1964 the main trials of Nuffield material in schools began, involving seventy-three schools and nearly 150 teachers. The schools were grouped geographically, and for each group there was an area leader responsible for teaching and supply problems, and for acting as a link between individual schools and the Headquarters Team. With a substantial sum of money given by Imperial Chemical Industries Ltd, the Foundation supplied schools with the apparatus needed to carry out the trials effectively.

Using information received during school trials, the Headquarters Team produced more written material for the *Teachers' guide*, which was tried in fifty-six schools during the academic year 1964–65.

Following consultations begun in February 1964 with the eight examining boards for the General Certificate in Education, a decision was taken to provide special alternative O-level chemistry examinations. These are now firmly established and are administered by the University of London School Examinations Department on behalf of all the GCE boards.

Many people have contributed to the chemistry teaching programme, first published in 1966. Over 200 teachers in schools of all types, and upwards of 5000 pupils, took part in a combined operation which was at the time unique in the educational history of the United Kingdom.

Since they were first published, the Nuffield proposals have been widely adopted, and considerable interest in them has been shown both in this country and overseas. By 1972 the total number of pupils taking Nuffield Chemistry at ordinary level had risen to 15 000, about 15 per cent of the total United Kingdom entry. Adaptations of parts of the Sample Scheme are being carried out in East Africa and Malaysia, affecting large numbers of pupils there. The major part of Stage I of the chemistry proposals has been adopted as the chemistry required for the Public Schools Common Entrance examination.

The second edition

By 1971 the Nuffield materials had been well tried in many schools. Dr Richard Ingle was appointed full-time to direct the preparation of a second edition of the materials. An advisory committee, consisting of teachers who have used Nuffield Chemistry in their schools, and two members of the original Headquarters Team, was assembled and much advice collected from many teachers in schools. This volume is one of the results. Fuller details of the revision are given in Part 1 of this book.

INDEX